Power Electronics

TUTORIAL GUIDES IN ELECTRONIC ENGINEERING

Series editors
Professor G.G. Bloodworth, *University of York*
Professor A.P. Dorey, *University of Lancaster*
Professor J.K. Fidler, *University of York*

This series is aimed at first- and second-year undergraduate courses. Each text is complete in itself, although linked with others in the series. Where possible, the trend towards a 'systems' approach is acknowledged, but classical fundamental areas of study have not been excluded. Worked examples feature prominently and indicate, where appropriate, a number of approaches to the same problem.

A format providing marginal notes has been adopted to allow the authors to include ideas and material to support the main text. These notes include references to standard mainstream texts and commentary on the applicability of solution methods, aimed particularly at covering points normally found difficult. Graded problems are provided at the end of each chapter, with answers at the end of the book.

1. Transistor Circuit Techniques: discrete and integrated (2nd edition) — G.J. Ritchie
2. Feedback Circuits and Op Amps — D.H. Horrocks
3. Pascal for Electronic Engineers — J. Attikiouzel
4. Computers and Microprocessors: components and systems — A.C. Downton (3rd edition)
5. Telecommunication Principles — J.J. O'Reilly
6. Digital Logic Techniques: principles and practice (2nd edition) — T.J. Stonham
7. Instrumentation Transducers and Interfacing — B.R. Bannister and D.G. Whitehead
8. Signals and Systems: models and behaviour — M.L. Meade and C.R. Dillon
9. Electromagnetism for Electronic Engineers — R.G. Carter (2nd edition)
10. Power Electronics — D.A. Bradley (2nd edition)
11. Semiconductor Devices: how they work — J.J. Sparkes
12. Electronic Components and Technology: engineering applications — S.J. Sangwine
13. Control Engineering — C. Bissell
14. Basic Mathematics for Electronic Engineers: models and applications — J.E. Szymanski
15. Integrated Circuit Design Technology — M.J. Morant

Power Electronics

Second edition

D.A. Bradley

Engineering Department
School of Engineering, Computing and Mathematics
Lancaster University, UK

CRC Press
Taylor & Francis Group
Boca Raton   London   New York

CRC Press is an imprint of the
Taylor & Francis Group, an informa business

CRC Press
6000 Broken Sound Parkway, NW
Suite 300, Boca Raton, FL 33487

270 Madison Avenue
New York, NY 10016

2 Park Square, Milton Park
Abingdon, Oxon OX14 4RN, UK

© 1987, 1995 D.A. Bradley Reprinted 2009 by CRC Press

Typeset in 10/12pt Times by Compuscript Ltd, Co. Clare, Ireland
Printed in England by Clays Ltd, St Ives plc.

ISBN 0 412 57100 5

A catalogue record for this book is available from the British Library

Library of Congress Catalog Card Number: 94-71066

∞ Printed on acid-free text paper, manufactured in accordance with
ANSI/NISO Z39.48-1992 and ANSI/NISO Z39.48-1984
(Permanence of Paper).

Contents

Preface

The subject of power electronics originated in the early part of the twentieth century with the development and application of devices such as the mercury arc rectifier and the thyratron valve. Indeed many of the circuits currently in use and described in this book were developed in that period. However, the range of applications for these early devices tended to be restricted by virtue of their size and problems of reliability and control.

With the development of power semiconductor devices, offering high reliability in a relatively compact form, power electronics began to expand its range and scope, with applications such as DC motor control and power supplies taking the lead. Initially, power semiconductor devices were available with only relatively low power levels and switching speeds. However, developments in device technology resulted in a rapid improvement in performance, accompanied by a corresponding increase in applications. These now range from power supplies using a single power MOSFET to high voltage DC transmission where the mercury arc valve was replaced in the 1970s by a solid-state 'valve' using thyristor stacks.

Developments in microprocessor technology have also influenced the development of power electronics. This is particularly apparent in the areas of control, where analogue controllers have largely been replaced by digital systems, and in the evolution of the 'smart power' devices. These developments have in turn led to system improvements in areas such as robot drives, power supplies and railway traction systems.

To a professional electrical engineer, power electronics encompasses all of the above, from the mercury arc rectifier systems still operational to the microprocessor-controlled drives on a robot arm. However, to deal with all of these topics is outside the scope of this book which concentrates on providing the reader with an introduction to the subject of power electronics. Following a discussion of the major power electronic devices and their characteristics, with relatively little consideration given to device physics, the emphasis is placed on the systems aspects of power electronics and on the range and diversity of applications. For this reason, a number of 'mini case studies' are included in the chapters on applications. These case studies cover topics from high-voltage DC transmission to the development of a controller for domestic appliances such as washing machines and are intended to place the material under discussion into a practical context.

As the text is intended for instruction and learning rather than for reference, each chapter includes a number of worked examples for emphasis and reinforcement. These worked examples are supported further by a number of exercises at the end of each chapter.

The production of a book of this type does not proceed in isolation and many people have contributed to it, either by the provision of material or with encouragement and advice. From among my colleagues at Lancaster

University I would particularly like to thank Professor Tony Dorey, who encouraged me to undertake the project and in the first instance provided much helpful advice.

As this book is one of a series of related texts, its relationship with the other texts in the series is important. I am therefore grateful to Professor John Sparkes of the Open University for his help in this respect, particularly with Chapter 1.

I would also like to thank Mr. A. Woodworth of Mullard, Mr. J.M.W. Whiting of GEC Traction, Mr. I.E. Barker of GEC Power Transmission and Distribution Projects, Mr. T.G. Carthy, Mr. A. Polkinghorne and Professor P. McEwan, all of whom helped with the provision of material.

As the book is intended for students, a student's viewpoint at an early stage proved particularly valuable and I repeat my thanks to Michael Anson for having diverted himself from his studies for a time to help me in that respect. Finally, but by no means least, my thanks go to Professor G. Bloodworth of York University for his efforts, constructive comments and assistance in this project.

1
Power semiconductors

□ To introduce the major power electronic devices.
□ To define their operating regimes and modes of operation.
□ To establish their ratings.
□ To consider losses and heat transfer properties.
□ To examine means of protection.

Objectives

An intrinsic semiconductor is defined as being a material having a resistivity which lies between that of insulators and conductors and which decreases with increasing temperature. The principal semiconductor material used for power electronic devices is silicon, a member of Group IV of the periodic table elements which means it has four electrons in its outer orbit.

If an element of Group V, such as phosphorus, with five electrons in its outer orbit is added to the silicon, each phosphorus atom forms a covalent bond within the silicon lattice, leaving a loosely bound electron. The presence of these additional electrons greatly increases the conductivity of the silicon and a material doped in this way is referred to as an n-type semiconductor.

By introducing an element from Group III as impurity, a vacant bonding location or hole is introduced into the lattice. This hole may be considered to be mobile as it can be filled by an adjacent electron, which in its turn leaves a hole behind. Holes can be thought of as carriers of positive charge and a semiconductor doped by a Group III impurity is referred to as a p-type semiconductor.

The extra, mobile electrons introduced by doping into the n-type material and the equivalent holes in the p-type material are referred to as the majority carriers. In an n-type material there is a small population of holes and in a p-type material a small population of electrons. These are called the minority carriers.

Diode

The semiconductor junction diode shown in Fig. 1.1 is the simplest semiconductor device used in power electronics. With no external applied voltage the redistribution of charges in the region of the junction between the p-type and n-type materials results in an equilibrium condition in which a potential

Ghandi, S.K. (1977). *Semiconductor Power Devices.* Wiley Interscience.

Sparkes, J. *Semiconductor Devices.* (1987). Van Nostrand Reinhold.

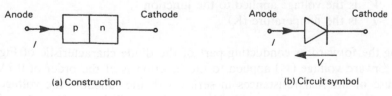

(a) Construction (b) Circuit symbol

Fig. 1.1 The diode.

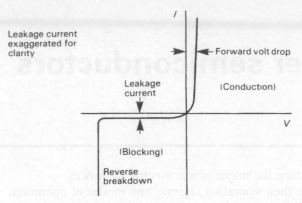

Fig. 1.2 Diagrammatic representation of the diode static characteristic. (Note: forward and reverse voltage scales are unequal. The forward voltage drop is of the order of 1 V while the reverse breakdown voltage varies from a few tens of volts to several thousand.)

The region over which the potential barrier exists is known as the depletion or transition layer.

The magnitude of the reverse leakage current can vary from a few picoamperes for an integrated circuit diode to a few milliamperes for a power diode capable of carrying several thousand amperes in the forward direction.

Avalanche diodes are designed to operate safely under reverse breakdown conditions and are used for device overvoltage protection (section on protection, p. 32).

barrier is established across a narrow region depleted of charge carriers on each side of the junction. This equilibrium may be disturbed by the application of an external applied voltage of either polarity.

If a reverse voltage – cathode positive with respect to anode – is applied, the electric field at the junction is reinforced, increasing the height of the potential barrier and increasing the energy required by a majority carrier to cross this barrier. The resulting small reverse leakage current shown in the diode static characteristic of Fig. 1.2 is due to the flow of minority carriers across the junction. The magnitude of the reverse leakage current increases with temperature because the number of minority carriers available increases with the temperature of the material.

The reverse current will be maintained with increasing reverse voltage up to the point at which reverse breakdown occurs, which will not cause the destruction of the diode unless accompanied by excessive heat generation.

When a forward voltage – anode positive with respect to cathode – is applied to the diode the height of the potential barrier is reduced, giving rise to a forward current resulting from the flow of majority carriers across the junction. As the forward voltage is increased, the forward current through the diode increases exponentially. The overall current–voltage characteristic of the diode is given approximately by

$$I = I_s \left[\exp(qV_j/kT) - 1 \right] \tag{1.1}$$

where I_s is the reverse leakage current
q is the electronic charge (1.602×10^{-19} C)
k is Boltzmann's constant (1.38×10^{-23} J K^{-1})
V_j is the voltage applied to the junction
and T is the temperature (K)

giving the forward or conducting part of the diode characteristic of Fig. 1.2. The forward voltage (V_j) applied to the junction is of the order of 0.7 V but because of internal resistances in series with the junction, the voltage (V) across the terminals of a practical power diode will be of the order of 1 V,

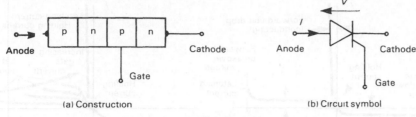

(a) Construction   (b) Circuit symbol

Fig. 1.3 The thyristor.

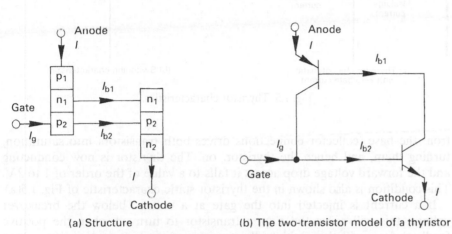

(a) Structure   (b) The two-transistor model of a thyristor

Fig. 1.4 The two-transistor model of a thyristor.

with the actual value being determined by the magnitude of the forward current and device temperature.

Thyristor

The thyristor is a four-layer, three-terminal device as suggested by Fig. 1.3. The complex interactions between three internal p–n junctions are then responsible for the device characteristics. However, the operation of the thyristor and the effect of the gate in controlling turn-on can be illustrated and followed by reference to the *two-transistor model* of Fig. 1.4. Here, the p_1–n_1–p_2 layers are seen to make up a p–n–p transistor and the n_2–p_2–n_1 layers an n–p–n transistor with the collector of each transistor connected to the base of the other.

With a reverse voltage, cathode positive with respect to the anode, applied to the thyristor the p_1–n_1 and p_2–n_2 junctions are reverse biased and the resulting characteristic is similar to that of the diode with a small reverse leakage current flowing up to the point of reverse breakdown as shown by Fig. 1.5(a).

With a forward voltage, anode positive with respect to the cathode, applied and no gate current, the thyristor is in the *forward blocking* mode. The emitters of the two transistors are now forward biased and no conduction occurs. As the applied voltage is increased, the leakage current through the transistors increases to the point at which the positive feedback resulting

Taylor, P.D. (1987). *Thyristor Design and Realisation.* John Wiley.

Richie, G.J. (1983). *Transistor Circuit Techniques,* Van Nostrand Reinhold gives a discussion of transistor types.

Silicon Controlled Rectifier Manual. General Electric (1979).

Power Semiconductor Handbook. Semikron (1980).

3

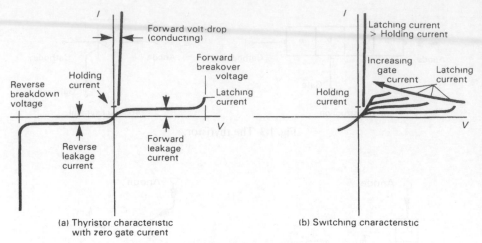

| | |
| (a) Thyristor characteristic with zero gate current | (b) Switching characteristic |

Fig. 1.5 Thyristor characteristics.

This is the forward breakover condition.

from the base/collector connections drives both transistors into saturation, turning them, and hence the thyristor, on. The thyristor is now conducting and the forward voltage drop across it falls to a value of the order of 1 to 2 V. This condition is also shown in the thyristor static characteristic of Fig. 1.5(a).

The latching current is the minimum current required to ensure conduction is maintained.

If a current is injected into the gate at a voltage below the breakover voltage, this will cause the n–p–n transistor to turn on when the positive feedback loop will then initiate the turn on of the p–n–p transistor. Once both transistors are on, the gate current can be removed because the action of the positive feedback loop will be to hold both transistors, and hence the thyristor, in the on state. The effect of the gate current is therefore to reduce the effective voltage at which forward breakover occurs, as illustrated by Fig. 1.5(b).

The holding current is less than the latching current.

Once conduction has been established in the thyristor the gate current may be reduced to zero.

Once the thyristor has been turned on it will continue to conduct as long as the forward current remains above the holding current level, irrespective of gate current or circuit conditions. Figures 1.6(a) and 1.6(b) show a single, ideal thyristor supplying a resistive and an inductive load respectively. In each case the thyristor is being turned on after a delay of about a quarter of a cycle after the voltage zero. In the case of the resitive load the load current follows exactly the load voltage. However, in the case of the inductive load the load voltage is made up of two components, the voltage across the inductance (v_i) and the voltage across the resistance (v_r). The current through the thyristor has an initial value of zero. The current then rises to a maximum at which point di/dt and hence the voltage across the inductance (v_i) becomes zero and the load voltage (v_L) equals the voltage across the resistor (v_r). The slope of di/dt then becomes negative, changing the polarity of v_i, and thus maintains the forward voltage drop across the thyristor until the stored energy in the inductance has been dissipated.

Turn-on

Following the initiation of forward breakover by the gate current the process of establishing conduction is independent of the gate conditions once the

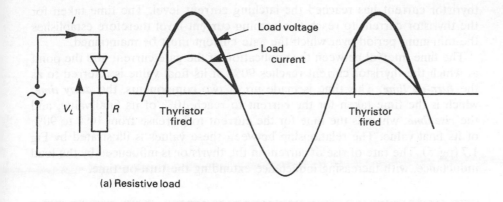

(a) Resistive load

$v_l = v_L - v_r$

$v_r = iR$
$v_l = L\, di/dt$

Load voltage (v_L)

v_r

Load current (i)

$di/dt + ve$
$di/dt - ve$

Thyristor fired

Thyristor fired

(b) Inductive load

Fig. 1.6 Thyristor with different loads.

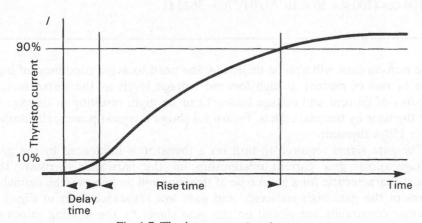

90%

10%

Delay time

Rise time

Time

Fig. 1.7 Thyristor current on turn-on.

Instantaneous power=Voltage
across thyristor × Current
through thyristor.

thyristor current has reached the latching current level. The time taken for the thyristor current to reach the latching current level therefore establishes the minimum period over which the gate current must be maintained.

The time interval between the application of the gate current and the point at which the thyristor current reaches 90% of its final value is referred to as the *turn-on time*. This time is made up of two components, the *delay time*, which is the time taken for the current to reach 10% of its final value, and the *rise time*, which is the time for the current to increase from 10% to 90% of its final value. The relationship between these values is illustrated by Fig 1.7 (pg. 5). The rate of rise of current in the thyristor is influenced by the load inductance, with increasing inductance extending the turn-on time.

Worked example 1.1

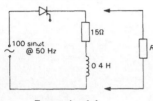

Example 1.1

A thyristor with a latching current of 40 mA is used in the circuit shown. If a firing pulse of 50 μs is applied at the instant of maximum source voltage, show that the thyristor will not be turned on. What value of resistance R' connected as shown will ensure turn-on?

Following the application of the gate pulse
$$100 \cos \omega t = iR + L\,di/dt$$
Using Laplace transforms
$$i = 100\,[\cos(\omega t - \phi) - \cos\phi.\exp(-Rt/L)]/(R^2 + \omega^2 L^2)^{1/2}$$
$$\phi = \tan^{-1}\omega L/R = 83.19° = 1.452\,\text{rad}$$
$$(R^2 + \omega^2 L^2)^{1/2} = 126.6$$
After 50 μs, by substituting in equation for i
$$i = 0.0124\,\text{A}$$
Hence thyristor fails to turn on.
Connecting R' then current in R' is i' when
  Current in thyristor $= i_t = i + i'$
For turn-on
$$i + i = 0.04\,\text{A}$$
$$\therefore\ i' = 40 - 12.4 = 27.6\,\text{mA}$$
Maximum value of R' is thus
$$100 \cos (100\pi \times 50 \times 10^{-6})/0.0276 = 3623\ \Omega$$

The turn-on time will also be limited by the need to avoid conditions of high rate of rise of current at high forward voltage levels as the instantaneous product of current and voltage (power) can be high, resulting in damage to the thyristor by thermal effects. Figure 1.8 shows a typical power relationship for a 150 A thyristor.

The gate resistance curve.

Thyristor gate current and
voltage.

The gate signal required to turn on a thyristor is influenced by the gate voltage versus gate current relationship for the particular thyristor. The actual characteristic for a given type of thyristor will lie between the definable limits of the gate high resistance and gate low resistance lines of Fig. 1.9. Further constraints are placed on the gate signal by the limiting values of gate current, gate voltage, gate power and temperature. Combining these limits gives the full gate characteristic of Fig. 1.9.

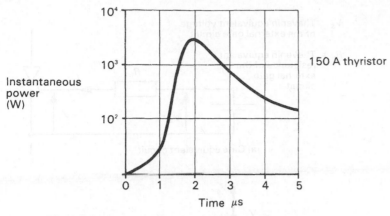

150 A thyristor

Instantaneous
power
(W)

Time μs

Fig. 1.8 Instantaneous power in thyristor during turn-on.

Firing of a thyristor by a
pulse train.

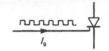

The actual operating point is obtained from consideration of the circuit of Fig. 1.10. This circuit defines the gate load line (slope $= -R_G$), the intersection of which with the thyristor gate resistance characteristic determines the gate operating point.

Typically, thyristor firing circuits use pulse techniques which allow a precise control of the point-on-wave at which the thyristor is fired and which dissipate less energy in the gate than a continuous current. Reliance is not usually placed on a single pulse to fire the thyristor but instead the firing circuit is arranged to generate a train of pulses.

The connection of the gate signal to a thyristor often makes use of a pulse transformer as in Fig. 1.11(a) to provide isolation and remove the need for a floating gate power supply. The circuit of Fig. 1.11(a) may be simplified to that of Fig. 1.11(b) in which case, the effective voltage applied to the gate can be shown to be

$$V_g = \frac{1}{n} \frac{V_s R'_g}{(R_s + R'_g)} \exp(-t/\tau)$$ (1.2)

Pulse transformers are typically constructed around a low-loss core and characteristically have negligible leakage inductance and a relatively high winding resistance.

R'_g in Fig. 1.11(b) is the effective impedance of the gate circuit transferred across the pulse transformer.

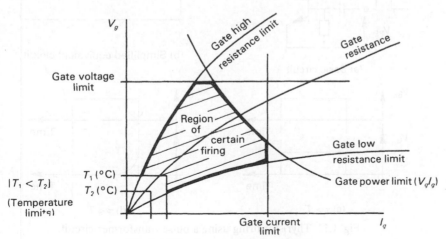

Fig. 1.9 Thyristor gate characteristic.

7

V_S = Thevenin equivalent voltage
of the external gate circuit

R_G = Thevenin equivalent
resistance of the
external gate
circuit

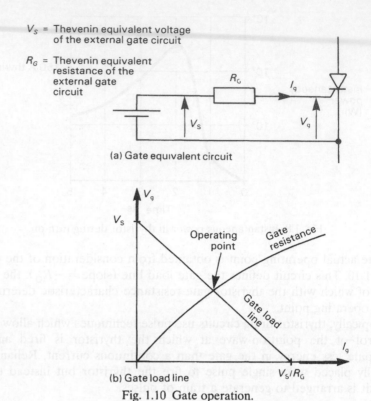

(a) Gate equivalent circuit

(b) Gate load line

Fig. 1.10 Gate operation.

The actual operating point can be read from consideration of the circuit of Fig. 1.10. This circuit defines a linear load line (slope = $-1/R_G$) the intersection of which with the thyristor gate resistance characteristic determines the gate operating point.

Typically, thyristor gate circuits use these continuous which allows positive control of the point to wave at which the thyristor is fired and which dissipation is greatest in the assertion instantaneous current. Reliance is not usually placed on a single pulse from the circuit but instead the firing circuit is arranged to generate a train...

The connection of the gate circuit to a thyristor often makes use of a pulse transformer as in Fig. 1.11(a) to provide isolation and remove the need for a floating gate power supply. The circuit of Fig. 1.11(a) can be simplified to that of Fig. 1.11(b) in which case the resistive voltage developed by the pulse can be shown to be...

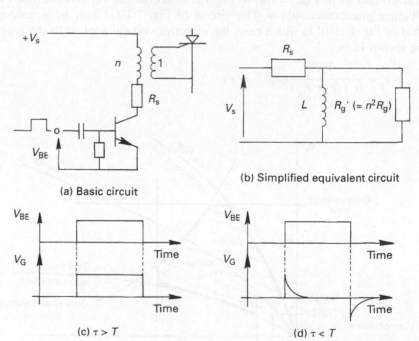

(a) Basic circuit

(b) Simplified equivalent circuit

(c) $\tau > T$

(d) $\tau < T$

Fig. 1.11 Thyristor firing using a pulse transformer circuit.

where n is the turns ratio of the pulse transformer
and

$$R'_g = n^2 R_g$$

where R_g is the gate circuit resistance of the thyristor

$$\tau = \frac{R_e}{L} \text{ and } R_e = \frac{R_s R'_g}{(R_s + R'_g)}$$

For those cases where the system time constant (τ) is much greater than the pulse width T the pulse will be correctly transmitted as in Fig. 1.11(c). However, if the system time constant is much less than the pulse width the response will be as shown in Fig. 1.11(d). This has the advantage that the resulting gate waveform enables the injection of a large initial charge without significant heating of the gate permitting a high $\mathrm{d}i/\mathrm{d}t$ value in the thyristor immediately after firing.

Isolation is also required when the gate signal is derived from a microprocessor in which case an opto-isolated device can be used in the gate circuit. Indeed, it is possible to obtain thyristors with integral opto-isolation. Such thyristors tend to be relatively low power devices but can be used to provide isolation in the gate circuit of a larger device as in Fig. 1.12(a). Figure 1.12(b) shows an alternative approach using an optically coupled transistor.

Typically $\tau > 10T$.

Typically $\tau < T/10$.
Where the same pulse is to be used to fire multiple devices, the value of V_S must be chosen to ensure the firing of the least sensitive of these devices.

Turn-off

Turn-off of a thyristor begins when the forward current falls below the holding current level of Fig. 1.5(a) with no gate current applied. Turn-off performance depends on device characteristics, the forward current prior to turn-off, the peak reverse current and the rate of rise of forward voltage as

Details of forced commutation circuits are given in Chapter 3.

(a) Thyristor

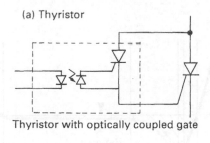

Thyristor with optically coupled gate

(b) Transistor

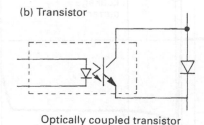

Optically coupled transistor

Fig. 1.12 Gate isolation using optically coupled devices.

well as temperature effects. Once the current has fallen to zero the thyristor must be placed into the reverse blocking state with a reverse voltage applied across the thyristor for sufficient time to allow the potential barriers to be re-established, completing the turn-off.

The dynamic behaviour of the thyristor during turn-off is shown in Fig. 1.13. Initially, the forward current falls, reaching zero at time t_0 and then reverses. From t_0 to t_1 the reverse current is sustained by the large numbers of carriers previously injected into the thyristor and device voltage drop is small. The build-up of the potential barriers at the junctions between and the removal of charge carriers by the action of the reverse current in the interval from t_1 to t_2 means that at time t_2 the reverse current can no longer be sustained and it begins to reduce. At this point the full reverse voltage appears across the junction, and as the circuit is slightly inductive this voltage will overshoot slightly, driving the reverse current down to the level of the reverse leakage current.

This occurs with all p-n junctions when changing from a forward biased to a reverse biased condition. (Sparkes, J. *Semiconductor Devices* (1987). Van Nostrand Reinhold.)

The carrier stored charge recovered during this period is shown as the shaded area of Fig. 1.13 and is referred to as the ***reverse recovery charge*** (Q_{rr}). Although the reverse recovery period is completed at time t_3, the reverse voltage must be maintained until time t_4 to ensure that the carrier density in the region of the central junction is reduced to a sufficiently small level to prevent the possibility of turn-on occurring when a forward voltage is reapplied. The total time for turn-off will vary according to the thyristor but will typically lie in the range 10 to 100 μs.

These conditions for turn-off occur automatically in a naturally commutated converter such as those described in Chapter 2. There is, however, a range of circuits operating from a DC voltage source in which additional circuitry must be used to turn the thyristor off. These additional, forced commutation circuits first force a reverse current through the thyristor for a short time to reduce the forward current to zero and then maintain the reverse voltage for the necessary time interval to complete the turn-off.

Details of forced commutation circuits are given in Chapter 3.

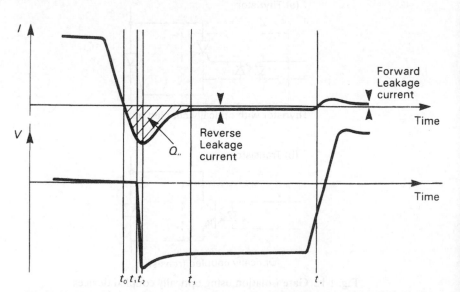

Fig. 1.13 Thyristor current and voltage during turn-off with zero gate current.

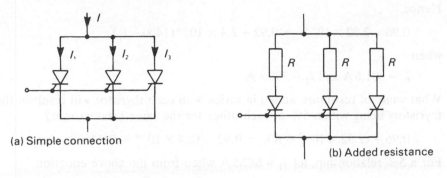

(a) Simple connection

(b) Added resistance

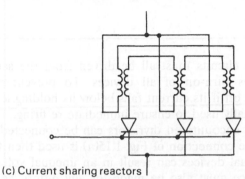

(c) Current sharing reactors

Fig. 1.14 Parallel connection of thyristors.

Parallel and series operation of thyristors

To accommodate high load currents a parallel connection of thyristors can be used. If the simple connection of Fig. 1.14(a) is used then differences in the individual thyristors will result in an unequal sharing of current between them. This sharing can be evened out by careful selection of matched devices, by the use of series resistance, as in Fig. 1.14(b), or by including current-sharing reactors as in Fig. 1.14(c).

Two thyristors are connected as in the margin figure and have forward characteristics when conducting of the form

| Thyristor 1 | $v = 0.96 + 2.52 \times 10^{-4} i$ |
| Thyristor 2 | $v = 0.92 + 2.4 \times 10^{-4} i$ |

What will be the current distribution between the two thyristors when the total current is 1400 A?

$$v = 0.96 + 2.52 \times 10^{-4} i_1 = 0.92 + 2.4 \times 10^{-4} i_2$$

and

$$i_1 + i_2 = 1400$$

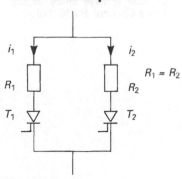

Example 1.2

Hence

$$0.96 + 2.52 \times 10^{-4} i_1 = 0.92 + 2.4 \times 10^{-4}(1400 - i_1)$$

when

$$i_1 = 601.6 \,\text{A and } i_2 = 798.4 \,\text{A}$$

What value of resistance added in series with each thyristor will result in the thyristors being within 5% of each other for the same total current?

$$0.96 + (2.52 \times 10^{-4} + R)i_1 = 0.92 + (2.4 \times 10^{-4} + R)(1400 - i_1)$$

For a 5% relationship, let $i_1 = 682.5 \,\text{A}$ when from the above equation

$$R = 1.14 \,\text{m}\Omega$$

At turn-on the thyristor gate circuits must all be driven from the same source to force a simultaneous turn-on of all devices. To prevent any individual thyristor from turning off if its current falls below its holding level a continuous gate signal is normally used to ensure immediate re-firing.

Where high voltage levels are encountered thyristors can be connected in series to share the voltage. If the connection of Fig. 1.15(a) is used then the differences between the individual devices can result in an unequal voltage sharing between them. Allowance must also be made for any difference in recovery times to ensure that all thyristors are left able to withstand the reapplication of forward voltage.

Voltage sharing can be achieved by using equalization networks such as that of Fig. 1.15(b). The capacitors ensure that each thyristor recovers fully on turn-off, while the resistor R_1 prevents an excessive $\mathrm{d}i/\mathrm{d}t$ on turn-on and resistor R_2 provides for equal steady-state sharing of voltage.

As the gates of the individual thyristors can be separated by potentials of several thousand volts, a gate pulse must be provided from a common source via some means of isolation such as transformers, optically coupled diodes or transistors and fibre-optic light guides.

Gate circuit isolation.

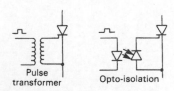

Pulse transformer

Opto-isolation

Chapter 5, case study of the Cross-Channel HVDC link.

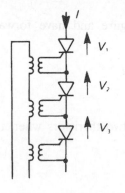

(a) Simple series connection

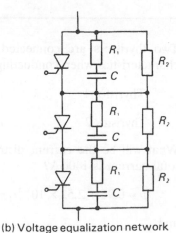

(b) Voltage equalization network

Fig. 1.15 Series connection of thyristors.

The operation of all power semiconductors is limited by a series of ratings which define the operating boundaries of the device. These ratings include limits on the peak, average and RMS currents, the peak forward and reverse voltages for the devices, maximum rates of change of device current and voltage, device junction temperature and, in the case of the thyristor, the gate current limits.

The current ratings of a power semiconductor are related to the energy dissipation in the device and hence the device junction temperature. The maximum value of *on-state current* ($I_{av(max)}$) is the maximum continuous current the device can sustain under defined conditions of voltage and current waveform without exceeding the permitted temperature rise in the device. The *RMS current rating* (I_{RMS}) is similarly related to the permitted temperature rise when operating into a regular duty cycle load such as that of Fig. 1.16.

In the case of transient loads, as the internal losses and hence the temperature rise in a power semiconductor are related to the square of the device forward current, the relationship between the current and the permitted temperature rise can be defined in terms of an $\int i^2 dt$ rating for the device.

On turn-on, current is initially concentrated into a very small area of the device cross-section and the devices is therefore subject to a di/dt rating which sets a limit to the permitted rate of rise of forward current.

The voltage ratings of a power semiconductor device are primarily related to the maximum forward and reverse voltages that the device can sustain. Typically, values will be given for the *maximum continuous reverse voltage* ($V_{RC(max)}$), the *maximum repetitive reverse voltage* ($V_{RR(max)}$) and the *maximum transient reverse voltage* ($V_{RT(max)}$). Similar values exist for the forward voltage ratings.

The presence of a fast transient of forward voltage can cause a thyristor to turn on and a dv/dt rating is therefore specified for the device. The magnitude of the imposed dv/dt can be controlled by the use of a snubber circuit connected in parallel with the thyristor. Figure 1.17 shows the basic series *RC* snubber together with some more complicated variations.

Comparisons of power semiconductor device ratings are given in Fig. 1.49 and Table 1.2.

Different manufacturers may adopt slightly different formats to those given in the text for a number of these standard values.

Typically, inductance would be added in series with the load to limit di/dt where the load inductance by itself was insufficient.

$V_{RT(max)}$ is greater then $V_{RR(max)}$.

Equivalent forward voltage ratings are $V_{FC(max)}$, $V_{FR(max)}$ and $V_{FT(max)}$.

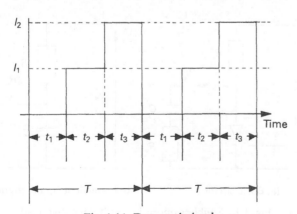

Fig. 1.16 Duty cycle load.

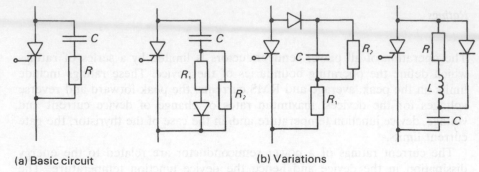

(a) Basic circuit   (b) Variations

Fig. 1.17 Snubber circuits.

Snubber circuits

Software is now available for the calculation and optimization of snubber circuits.

Consider the circuit of Fig. 1.18(a) in which a thyristor is being used to supply a resistive load. With the thyristor turned off, the effective load is as shown in Fig. 1.18(b). If a voltage step (V_s) is now applied the current through the capacitor becomes

$$i_S = \frac{V_S}{R}\exp(-t/CR) \tag{1.3}$$

The voltage across the capacitor is then

$$v_C = V_S(1 - \exp(-t/CR)) \tag{1.4}$$

when

$$\frac{dv_C}{dt} = \frac{V_S}{CR}\exp(-t/CR_S) \tag{1.5}$$

If the dv/dt rating of the thyristor is not to be exceeded then, from Equation 1.5,

$$C > \frac{V_S}{R\left[\dfrac{dv_C}{dt}\right]_{max}} \tag{1.6}$$

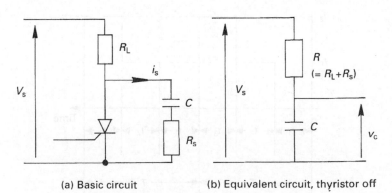

(a) Basic circuit   (b) Equivalent circuit, thyristor off

Fig. 1.18 Snubber operation with a resistive load.

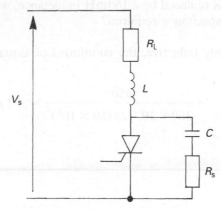

Fig. 1.19 Thyristor with inductive load.

If the load contains inductance as in Fig. 1.19, the value for dv/dt becomes

$$\frac{dv_C}{dt} = \frac{V_S \exp(-Rt/2L)}{\omega LC} \sin(\omega t) \qquad (1.7)$$

the maximum value of which is

$$\left[\frac{dv_C}{dt}\right]_{max} = \frac{V_S \exp[(-R/2\omega L)\tan^{-1}(2\omega L/R)]}{(LC)^{1/2}} \qquad (1.8)$$

where

$$\omega = \left[\frac{1}{LC} - \frac{R^2}{4L^2}\right]^{1/2}$$

For small values of inductance (say $L/R \leq 0.1$), the solution will revert to that of Equation 1.5 for a resistive load. For large values of inductance R/L tends to zero and ω to $(LC)^{-1/2}$, in which case

$$v_C = V_S(1 - \cos(\omega t)) \qquad (1.9)$$

then

$$C = \frac{V_S^2}{L\left[\dfrac{dv_C}{dt}\right]_{max}^2} \qquad (1.10)$$

Ignores the effect of snubber resistance voltage as this is usually small compared with the capacitor voltage.

Shepherd, W. and Hulley, L.N. (1987). *Power Electronics and Motor Control*. Cambridge University Press.

A thyristor has a maximum dv/dt rating of $60\,\text{V}\,\mu\text{s}^{-1}$ and is used to supply a resistive load of $8.4\,\Omega$. If the maximum step change in voltage anticipated is $450\,\text{V}$, what will be the minimal viable size of capacitor for a simpler RC snubber circuit?

Worked example 1.3

From Equation 1.6

$$C > \frac{V_S}{R_S\left[\dfrac{dv_C}{dt}\right]_{max}} = \frac{450}{8.4 \times 60 \times 10^6} = 890\,\text{nF}$$

15

If the resistive load is replaced by a 160 mH inductance, what will be the new minimum value of capacitance required?

As the load is entirely inductive, the conditions of Equation 1.10 apply, in which case:

$$C = \frac{V_s^2}{L\left[\dfrac{dv_C}{dt}\right]_{max}^2} = \frac{450^2}{160 \times 10^{-3} \times (60 \times 10^6)^2} = 352 \, pF$$

Thyristor construction

The active element of a typical high-power thyristor consists of a silicon wafer 0.2–0.3 mm thick and with a diameter of 80 mm or more. This wafer must be carried in a housing designed to provide the necessary connections to the anode, cathode and gate as well as being mechanically robust to allow the thyristor to be handled and mounted safely. It must also have suitable heat transfer properties for the purpose of device cooling.

Capsule construction is also referred to as a 'puck'.

The principal constructions used for medium to high power thyristors are the stud-base and capsule forms of Figs. 1.20(a) and 1.20(b), while for lower power applications the flat-pack assembly of Fig. 1.20(c) is often used.

Increasingly, low to medium power devices are available as multi-chip modules incorporating appropriate device isolation and gating. Such modules

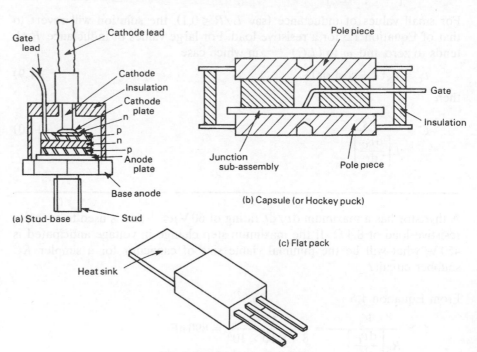

Fig. 1.20 Thyristor assemblies.

are increasingly used for applications such as pulse-width-modulated inverters where the module construction improves the integrity and robustness of the inverter.

Single devices with power ratings of a few watts are also available in surface mount packages, either as single devices or as modules containing a number of devices.

Gate turn-off thyristor

The *gate turn-off* (GTO) thyristor, the circuit symbol for which is shown in Fig. 1.21, is a variant of the thyristor in which the internal structure is modified to enable the forward current to be turned off by the application of a negative gate current. GTO thyristors are now used in a wide variety of applications requiring controlled device turn-off, including pulse-width modulated inverters.

The turn-on mechanism for a GTO thyristor is similar to that of a conventional thyristor, though in practice the gate current is maintained throughout the whole of the conduction period to prevent any drop out from the conducting state.

On turn-off, the GTO thyristor has a relatively low gain, of the order of 6 or 10, requiring the application of a high reverse current on the gate to force the device off.

The reverse voltage capability of a GTO thyristor may be less than that of an otherwise equivalently rated conventional thyristor. For this reason, a diode may need to be included in series with the GTO thyristor as in Fig. 1.22.

Asymmetric thyristor

The asymmetric thyristor is a combination of a thyristor and a reverse diode in parallel on the same wafer (Fig. 1.23). It is therefore controllable in the thyristor forward direction and will always conduct in the reverse direction. The asymmetric thyristor finds applications in circuits such as inverters where an inverse parallel diode is often employed in conjunction with a thyristor to

Chapter 3 gives a discussion of inverters.
Hollander, D. (1993). Discrete surface mount products for power applications. *Microelectronics Journal.* **24**, 9915-9.

Burgum, F.J. (1982). The GTO – a new power switch. *Electronics and Power*, May, 389-92.

The GTO thyristor is a fast switching device with turn-on and turn-off times of the order of $1\,\mu s$ achievable.

Hall, J.K. and Manning, C.D. (1984). Switching properties of GTO thyristors. IEE Conference Publication 234, *Power Electronics and Variable Speed Drives*, 958-61.

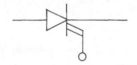

Fig. 1.21 Circuit symbol for a gate turn-off (GTO) thyristor.

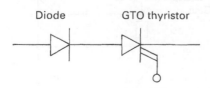

Fig. 1.22 GTO thyristor with series diode to improve reverse blocking performance.

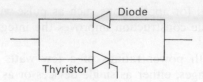

Fig. 1.23 Asymmetric thyristor.

accommodate the phase shift of current with an inductive load. By incorporating both devices on the same wafer they are in the closest possible proximity, reducing effects such as stray inductance which may otherwise influence performance.

Other thyristor types

Jayant Baliga, B. (1993). Power semiconductor devices for the 1900s. *Microelectronics Journal*, **24**, 31–9.
Hudgins, J.L. (1993). A review of modern power semiconductor devices. *Microelectronics Journal*, **24**, 41–54.

Switching times are of the order of 1 μs to 5 μs.

A number of other thyristor types are available though their applications so far tend to be relatively limited.

Static induction thyristor

The static induction thyristor (SITH) or field-controlled thyristor provides characteristics similar to those of a MOSFET. Turn-on is by means of a positive voltage applied to the gate while turn-off is achieved by the application of a negative gate voltage. A SITH provides fast switching with a low forward volt drop and high dv/dt and di/dt ratings. Devices with current and voltage ratings up to 500 A and 3 kV have been produced.

FET-controlled thyristors

A FET-controlled thyristor combines a MOSFET and thyristor as shown in Fig. 1.24. Turn-on is by the application of a voltage to the gate of the MOSFET but turn-off is achieved only through the reduction of the forward current below the holding current level of the thyristor. The FET-controlled thyristor offers a high switching speed at high values of dv/dt and di/dt.

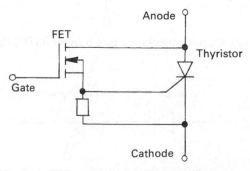

Fig. 1.24 FET-controlled thyristor configuration.

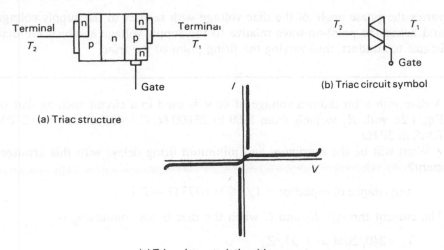

| Terminal T_2 | | Terminal T_1 |

(a) Triac structure

(b) Triac circuit symbol

(c) Triac characteristic with zero gate current

Fig. 1.25 The triac.

Triac

The triac is a multilayer device that is electrically the same as two thyristors connected in inverse parallel on the same wafer. Fig. 1.25 shows the make up of a triac together with its circuit symbol and static characteristic. As the terms anode and cathode have no real meaning for triacs, the terms **terminal 1** (T_1) and **terminal 2** (T_2) are used. The gate signal is then applied between the gate terminal and terminal 1. The triac can be turned on by the injection of either a positive or a negative gate current. It is, however, most sensitive when a positive gate current is used with terminal 2 at a positive potential (forward conduction) or a negative gate current with terminal 1 at a positive potential (reverse conduction). The least sensitive condition is a positive gate current applied with terminal 1 at a positive potential.

Triac firing. Typically a negative gate current would be used to initiate both forward and reverse conduction.

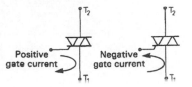

Diac

The diac is essentially a gateless triac configured to breakover at a low voltage in both the forward and reverse directions. Its main application is in trigger circuits such as that of Fig. 1.26 where the variation of resistance R_1

Diac circuit symbol and characteristic.

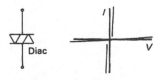

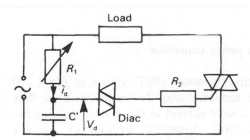

Fig. 1.26 Triac firing circuit using a diac.

varies the phase angle of the diac voltage with respect to the supply voltage and hence the point-on-wave relative to the supply voltage at which the diac begins to conduct, thus varying the firing point of the triac.

Worked example 1.4

A diac with a breakdown voltage of 40 V is used in a circuit such as that of Fig. 1.26, with R_1 variable from 1000 to 25 000 Ω, $C = 470$ nF, and $V = 240$ V RMS at 50 Hz.

What will be the maximum and minimum firing delays with this arrangement?

$$\text{Impedance of capacitor} = 1/\omega C = 6773\,\Omega = |Z_c|$$

The current through R_1 and C when the diac is not conducting is

$$i_d = 240\sqrt{2}\sin(\omega t + \phi)/Z_d$$

where

$$Z_d = \left(R_1^2 + 1/\omega^2 C^2\right)^{1/2}$$

and

$$\phi = \tan^{-1}(1/\omega RC)$$

with $R_1 = 1000\,\Omega$, $Z_d = 6846\,\Omega$
The voltage across the capacitor is $i_d Z_c = v_c$

$$v_c = 335.8\sin(\omega t - 8.4°)$$

When diac conducts, $v_c = 40$ V

$$\therefore \text{Minimum delay} = \sin^{-1}(40/335.8) + 8.4° = 15.24°$$

with $R_1 = 25\,000\,\Omega$, $Z_d = 25\,901\,\Omega$
The voltage across the capacitor is $i_d Z_c$ as before:

$$v_c = 88.76\sin(\omega t - 74.84°)$$

When diac conducts, $v_c = 40$ V
Hence

$$\text{Maximum delay} = \sin^{-1}(40/88.76) + 74.84° = 101.6°$$

Biopolar junction power transistor

Richie, G.J. (1988). *Transistor Circuit Techniques.* Van Nostrand Reinhold.

Blicher, A. (1981). *Field-effect and Bipolar Transistor Physics.* Academic Press.

The bipolar junction transistor (BJT) shown in Fig. 1.27(a) is a current-controlled three-layer semiconductor device in which the flow of current into the base controls the flow of current in the collector giving the forward characteristic of Fig. 1.28(a). Forward breakdown occurs with increasing collector–emitter voltage (V_{CE}) in a manner similar to that for previously discussed devices. However, reverse breakdown of the base emitter junction

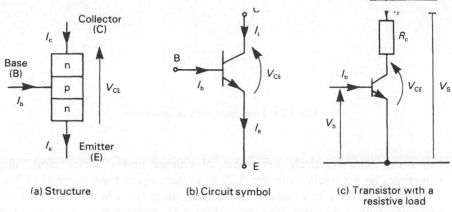

| (a) Structure | (b) Circuit symbol | (c) Transistor with a resistive load |

Fig. 1.27 N–p–n power transistor.

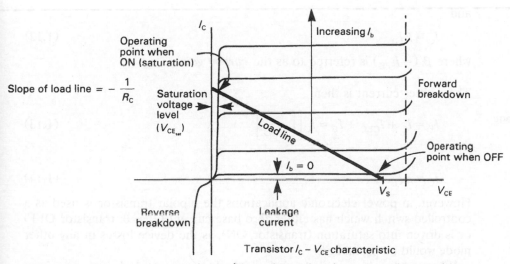

Fig. 1.28a Transistor characteristics. (Note: forward and reverse voltage scales are unequal.)

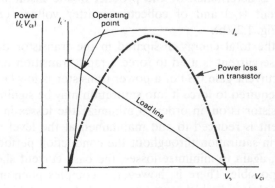

Fig. 1.28b Variation in power loss in a transistor with variation in V_{CE} for constant load R_C.

Fig. 1.29 Transistor with series diode.

The device is known as a bipolar transistor because both holes and electrons are active.

The base–emitter junction can fail at voltages as low as 10 V.

I_{CEO} is the collector leakage current and is highly temperature dependent.

$V_{CE,sat} \approx 1.1$ V

Power loss in the transistor is the product of collector current and collector–emitter voltage.

Energy = $\int$ Power $\cdot$ dt.

can occur at a relatively low value of voltage and a series diode may sometimes be necessary as in Fig. 1.29 to protect against the reversal of V_{CE}.

Conventionally, a transistor is used as an amplifier when for a transistor operating with a resistive load as in Fig. 1.27, operation is constrained to follow the load line shown on Fig. 1.28(a) since

$$I_C = (V_S - V_{CE})/R_C \tag{1.11}$$

and

$$I_C = \beta I_B \tag{1.12}$$

where β ($= h_{FE}$) is referred to as the *current gain*.

The emitter current is then

$$I_E = I_C + I_{CEO} + I_B \approx I_C\left[1 + \frac{1}{\beta}\right] \tag{1.13}$$

or

$$I_C \approx \alpha I_E \tag{1.14}$$

However, in power electronic applications the bipolar transistor is used as a controlled switch which has either zero base current ($I_B = 0$: transistor OFF) or is driven into saturation (transistor ON), as the device losses in any other mode would be prohibitive.

When used as a controlled switch, as the base current I_B is increased from 0 to the value required to drive the transistor into saturation the collector current increases while V_{CE} decreases. The instantaneous power loss during the transition is determined by the product of the instantaneous values of collector current (i_C) and of collector–emitter voltage (v_{CE}) and varies according to Fig. 1.28(b).

To reduce the total energy dissipated in the transistor during turn-on, a high initial base current is used to force a rapid transition into the saturated state. As the current gain (β) of a power transistor is low (≈ 10) the level of base current required to force it into saturation may be significant. Therefore, once the transistor is on, in order to minimize the losses in the base circuit the base current is reduced to and maintained at the level required to hold the transistor in saturation throughout the conduction period.

On turn-off, again to minimize losses, the base current should be reduced as rapidly as possible. There is, however, a complex phenomenon known as 'secondary breakdown' which can occur with fast switching transients and results in failure of the transistor which limits the rate at which the base

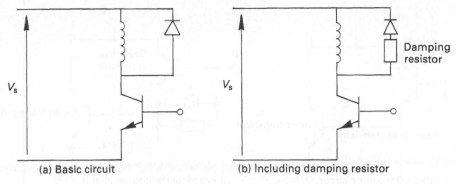

(a) Basic circuit   (b) Including damping resistor

Fig. 1.30 Transistor switching an inductive load.

current can be reduced. Finally, to improve turn-off a reverse base voltage may be applied and a subsequent reverse bias maintained while the device is off.

Despite the use of techniques to improve the turn-on and turn-off performance of the power transistor, the switching losses may well be the most significant form of loss in applications where high frequency switching is important such as pulse-width-modulated inverters or switched-mode power supplies.

Where a transistor is used to switch an inductive load, a reverse diode may be used as in Fig. 1.30(a) to dump the energy stored in the inductor and to prevent any excessive transient voltage appearing across the transistor. Where appropriate, a damping resistor may be used as in Fig. 1.30(b) to increase the rate at which the current in the inductive load falls off.

The safe operating area (SOA) of a transistor is defined in terms of the collector current and collector–emitter voltage for both pulsed and continuous (DC) operation as shown by Fig. 1.31. For safe operation, the instantaneous values of collector current (I_C) and the collector–emitter (V_{CE}) must

Chapter 3 discusses inverters and Chapter 6 discusses power supplies.
The same applies for any fast switching device operating with an inductive load.

Also referred to as the safe operating region (SOR).

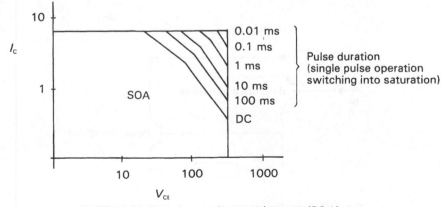

Fig. 1.31 Transistor safe operating area (SOA).

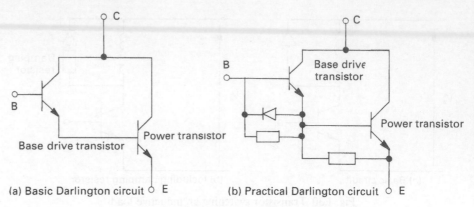

(a) Basic Darlington circuit   E   (b) Practical Darlington circuit   E

Fig. 1.32 Darlington transistor circuits.

lie within the boundaries of the appropriate curve throughout the switching interval.

Base isolation can be by means of a transformer or an optically coupled device.

Though a power transistor can switch more rapidly than a conventional thyristor, with times of the order of a microsecond achievable, the need to supply it with a continuous base current while in the ON state means that the requirements for the base drive circuit are more severe than for the thyristor gate circuit. Typically, a thyristor will require a relatively low energy pulse of a few milliseconds to turn it on, while an equivalently rated transistor requires a continuous current, often of a few amperes, to keep it turned on.

To overcome the problem of providing a large external base current, combinational devices such as the *Darlington pair* of Fig. 1.32 and the *complementary* form of Fig. 1.33 have been produced on a single chip. These arrangements reduce the demand on the external base drive circuits but at the expense of a significant reduction in the switching performance.

Worked example 1.5

A transistor has the switching characteristic shown in the figures. If the mean power loss in the transistor is limited to 200 W, what is the maximum switching rate that can be achieved?

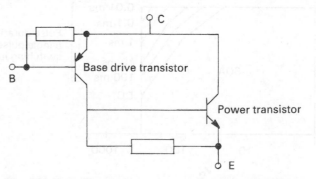

Fig. 1.33 Complimentary or super-α form of connection.

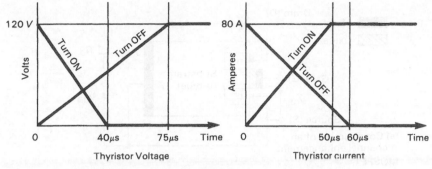

120 V

Volts

Turn ON   Turn OFF

0   40μs   75μs   Time

Thyristor Voltage

80 A

Amperes

Turn ON   Turn OFF

0   50μs 60μs   Time

Thyristor current

Example 1.4.

Energy loss in transistor $= \int_0^t i_c V_{CE} dt$

Turn-on energy loss $= \int_0^{40 \times 10^{-6}} 120(1 - 2.5 \times 10^4 t) \times 1.6 \times 10^6 t \, dt = 51 \, \text{mJ}$

Turn-off energy loss $= \int_0^{60 \times 10^{-6}} 1.6 \times 10^1 t \times 80(1 - 1.667 \times 10^4 t) dt = 76.8 \, \text{mJ}$

Total loss in one cycle $= 127.8 \, \text{mJ}$

Number of cycles in one second $= 200/0.1278 = 1564.9$

Power MOSFETs

Figure 1.34 shows the construction of an *n-channel* enhancement type MOSFET together with the circuit symbol for this and the *p-channel* form. The three terminals are referred to as the *gate, drain* and *source*.

A power MOSFET is a voltage-operated device requiring only a small gate current in order to turn it on and to maintain it in the ON state which means it is particularly suited to applications requiring high switching rates. However, as the gate circuit presents a capacitive load, with the result that there may be a significant reactive current in the gate circuit, the gate current source must be properly matched to the gate, especially where high speed switching is required.

The steady-state characteristic of a MOSFET is shown in Fig. 1.35. When operated in the linear region the drain current (I_D) varies proportionally to the drain–source voltage (V_{DS}). Pinch-off or saturation of the device occurs when $V_{DS} \geq (V_{GS} - V_T)$ when I_D remains essentially constant and independent of V_{DS}.

Operation in a switching mode is in the linear region of the characteristic in which case the output resistance (R_{DS}) is of the order of a few milliohms, rising to several megohms in the pinch-off region.

MOS = Metal Oxide Semiconductor; FET = Field Effect Transistor.

The gate current requirements for power MOSFETs are significantly less than the base current of equivalently rated power transistors.

For operation at low frequencies a series resistance is usually included in the gate circuit to limit any spurious high frequency currents.

V_T is the device threshold voltage below which no current will flow.

Output resistance is expressed as $R_{DS} = \Delta V_{DS}/\Delta I_D$

Insulated gate bipolar transistor

The insulated gate bipolar transistor (IGBT) of Fig. 1.36 combines a FET input stage in the gate together with a bipolar power stage. Turn-on is

25

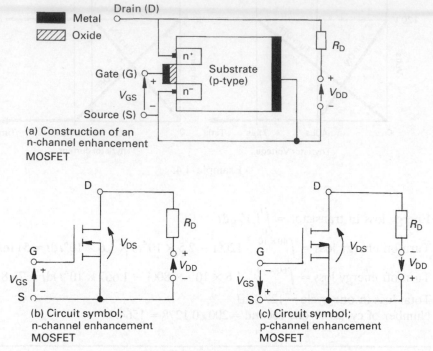

(a) Construction of an
n-channel enhancement
MOSFET

(b) Circuit symbol;
n-channel enhancement
MOSFET

(c) Circuit symbol;
p-channel enhancement
MOSFET

Fig. 1.34 Power MOSFETs.

achieved by the application of an appropriate gate–emitter voltage (V_{GE}). Turn-off occurs when the gate–emitter voltage is reduced to zero.

IGBTs are fast switching devices with turn-on times of better than 1 μs achievable. Turn-off times are generally longer than the turn-on time, though still of the order of 1 μs.

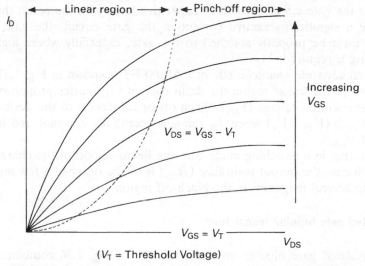

(V_T = Threshold Voltage)

Fig. 1.35 Output characteristics of an enhancement type MOSFET.

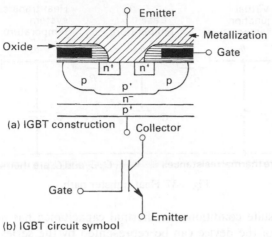

Emitter

Oxide → Metallization

← Gate

n⁺   n⁺

p   p⁺   p

n⁻

p⁺

(a) IGBT construction   Collector

Gate   Emitter

(b) IGBT circuit symbol

Fig. 1.36 Insulated gate bipolar transistor (IGBT).

IGBTs have found applications in low to medium power high-speed switching applications with devices with ratings of the order of 1200 V and 400 A available. On-state forward voltage drops range from 2 to 5 V dependent on current rating.

Heat transfer and cooling

The heat generated in a power semiconductor device due to internal losses has to be conducted away from the device and dissipated. In a thyristor, the internal heat source is taken to be a junction within the semiconductor material. The heat transfer path is then:

(a) from the junction to the case of the device;
(b) from the casing to a heat transfer system such as a fin;
(c) from the heat transfer system to the final ambient temperature heat sink.

The ambient temperature heat sink is taken as being a constant-temperature system whose temperature is not altered by any additional energy input.

The effect of the different thermal characteristics of these various stages in the heat removal process is analogous to a chain of resistors and capacitors fed from a current source. An equivalent circuit describing the thermal performance can therefore be formed using thermal resistance and capacitance as in Fig. 1.37.

Temperature differences are analogous to voltage differences and heat flow is analogous to current flow. This leads to an analogy between electrical and thermal components:

Electrical		Thermal	
Charge	$C = A\,s$	Heat	$J = W\,s$
Current	A	Heat flow	W
Potential difference	V	Temperature difference	$°C$
Resistance	$V\,A^{-1}$	Thermal resistance	$°C\,W^{-1}$
Capacitance	$A\,s\,V^{-1}$	Thermal capacitance	$W\,s\,°C^{-1}$

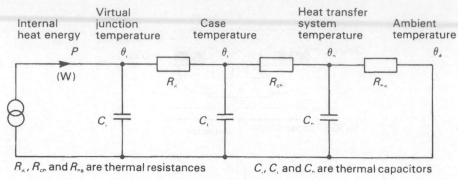

Fig. 1.37 Heat transfer path.

R_{jc}, R_{ch} and R_{ha} are thermal resistances C_j, C_c and C_h are thermal capacitors

Under steady-state conditions the thermal capacitance has no effect and the heat flow out of the device can be represented by the series combination of the thermal resistances:

Thermal analogue of Ohm's Law. Thermal resistance $R=$ (Temperature difference)/ (Power transferred)$=(\theta_1-\theta_2)/P$ (°C W^{-1}).

$$\theta_j = \theta_a + P(R_{jc} + R_{ch} + R_{ha}) \tag{1.15}$$

where
θ_j = Temperature of the junction (°C)
θ_a = Ambient temperature (°C)
P = Internal power loss (W)
R_{jc} = Thermal resistance, junction to casing (°C W^{-1})
R_{ch} = Thermal resistance, casing to heat transfer system (°C W^{-1})
R_{ha} = Thermal resistance, heat transfer system to heat sink (°C W^{-1})

The effect of using forced or liquid cooling is to increase the rate of heat removal, reducing R_{ha}.

Worked example 1.6

A thyristor has a thermal resistance of 0.82°C W^{-1} between its virtual junction and the heat transfer system and 1.96°C W^{-1} between the heat transfer system and the ambient temperature heat sink. What will be the permitted power loss in the thyristor if the junction temperature is to be kept below 132°C for an ambient temperature of 28°C?

From Equation 1.15,

$$132 = 28 + P(0.82 + 1.96)$$
$$\therefore \quad P = 37.4\,\text{W}$$

This illustrates the limitation of the analogy as impedance in electrical engineering is a steady-state term.

During transient conditions such as pulsed operation, overloads or faults, the temperature rise may be estimated by using a quantity referred to as the transient thermal impedance which incorporates the effects of thermal capacitance. The transient thermal impedance is a time-dependent function and relates the device temperature rise to the energy input in a defined time interval. For a sudden increase in dissipation at time $t = 0$ from 0 to P_{th}:

$$P_{th} Z_{th}(t) = \delta\theta(t) \tag{1.16}$$

where $Z_{th}(t)$ is the thermal impedance at time t
and $\delta\theta(t)$ is the temperature rise at time t.

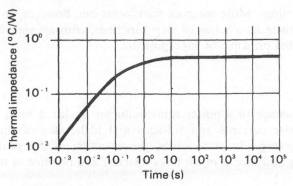

Fig. 1.38 Transient thermal impedance curve for a 100 A thyristor.

Curves of transient thermal impedance such as that shown in Fig. 1.38 are provided for individual semiconductors.

Consider a power semiconductor providing a continuous series of power pulses as in Fig. 1.39. The average junction temperature can be found from the mean power loss and the thermal impedance at $t = \infty$ as

The transient thermal impedance for $t = \infty$ is the thermal resistance.

$$\theta_j = \theta_a + P_{mean} Z_{ja}(\infty) = \theta_a + P_{mean} R_{ja} \qquad (1.17)$$

where θ_j is the junction temperature

θ_a is the ambient temperature

P_{mean} = mean power = $P_{max} t / T$

$Z_{ja}(\infty) = R_{ja}$ = thermal impedance at $t = \infty$

The junction temperature then varies about this mean by an amount $\delta\theta_j$, a good approximation to which can be obtained by considering the last two pulses transmitted when, referring to Fig. 1.39:

$$\delta\theta_j = [(P_{max} - P_{mean})Z_{th}(t_3)] - P_{max}[Z_{th}(t_2) - Z_{th}(t_1)] \qquad (1.18)$$

As the transient thermal impedance is an expression of the step response of the system it can be used directly only with waveforms that can be formed

Bird, B.M., King, K.G. and Pedder, D.A.G. (1993). *An Introduction to Power Electronics*. John Wiley, UK.

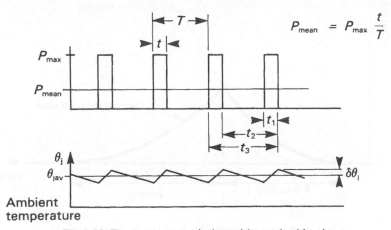

Fig. 1.39 Temperature variation with a pulsed load.

from step functions. More complex waveforms can, however, be examined by representing them as a series of step functions, with appropriate averaging, and applying the principle of superposition.

Protection

To prevent damage to a power semiconductor device it must be protected against excessive currents and voltages and high rates of change of both current and voltage. In many cases, the selection of the protection to be employed may well owe as much to experience and practice as to formulation and analysis.

Overcurrent protection

As the thyristor has a restricted overcurrent capacity, special fast-acting fuses are usually provided for overcurrent protection. These fuses must take account of:

(a) the need to permit the continuous passage of the steady-state current;
(b) permitted overload conditions including transients and duty-cycle loads;
(c) prospective fault conditions;
(d) the $i^2 t$ rating of the device – the fuse must clear the fault before the $i^2 t$ limit is reached;
(e) high peak-current levels during faults caused by current asymmetry;
(f) fuse voltage rating;
(g) ambient temperature conditions.

At high levels of fault current the fuse operation follows the pattern of Fig. 1.40 with operation within one cycle after time t_1. In the arcing phase, of duration t_2, the fuse is designed to present a high arc impedance in order to extinguish the arc and isolate the fault. The fuse characteristic during the

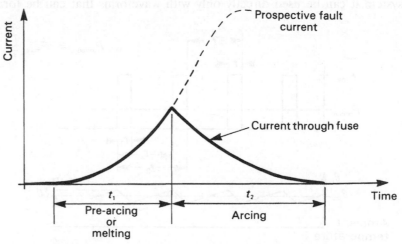

Fig. 1.40 Fuse behaviour with high levels of fault current.

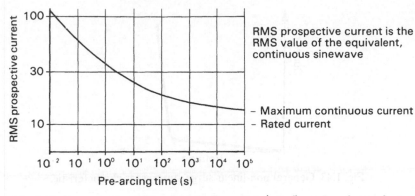

Fig. 1.41 Current–time characteristic for a 10 A (rated) semiconductor fuse.

melting phase is a function of the pre-arcing $\int i^2 dt$ characteristic of the fuse. Conditions once the fuse is open circuit are complicated by arc voltage and current conditions; however, the total pre-arcing and arcing $\int i^2 dt$ must be less than the $\int i^2 dt$ rating of the device being protected.

At lower fault currents, fuse melting occurs more slowly and hence less explosively with the arc extinguishing at a natural current zero following melting. The melting time for the fuse will become infinite at some particular current level. This is the maximum continuous current rating of the fuse and the ratio between this current and the rated current is known as the fusing factor and would normally be of the order of 1.3 to 1.4, as is illustrated by Fig. 1.41.

The overcurrent protection of transistors presents particular problems. A fault condition can cause an effective reduction in the transistor load (R_c in Fig. 1.26(c)), causing the load line of Fig. 1.27 to be rotated clockwise. As the supply voltage (V_s) is fixed, then for a fixed base current the transistor will come out of saturation, increasing the dissipation in the manner of Fig. 1.29. Though this increased dissipation may result in damage to the transistor, the rise in the collector current (I_c) may not be sufficient to blow a series fuse. To overcome this problem, the crowbar thyristor of Fig. 1.42 can be used. Here, additional circuitry is included to detect an increase in V_{CE} and fire the crowbar thyristor, causing the fuse to blow.

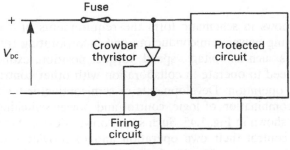

Fig. 1.42 Crowbar protection.

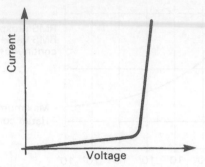

Fig. 1.43 General non-linear surge suppressor characteristic.

Overvoltage protection

Given that the thyristor is rated to withstand the normal maximum voltages expected in the system, the problem is that of protecting against transient overvoltages, which may also be accompanied by high levels of dv/dt.

Non-linear devices having characteristics of the type shown in Fig. 1.43 can be used in parallel with the active device in company with the reactive snubber circuits of Fig. 1.17 for overvoltage protection. Three typical non-linear devices are:

Voltage-dependent resistors or varistors. These are formed from a semiconductor material such as silicon carbide (SiC) or zinc oxide (ZnO) and can provide characteristics in which the current through the varistor varies as the thirteenth or higher power of applied voltage.

Avalanche diodes. These are diodes which can sustain large reverse currents in reverse breakdown without damage. The limitations on the use of an avalanche diode are the reverse breakdown voltage, the peak power loss and the steady-state power loss.

Selenium diode rectifiers. These are selemium−metal junction diodes constructed to have a low forward breakdown voltage and a well-defined and stable reverse breakdown voltage of around 72 V.

Smart power devices

Figure 1.44 shows in schematic form the requirements of a typical control system deploying power semiconductors and incorporating reference to system conditions such as voltage, speed, torque, position, extinction angle or current, the need to operate in collaboration with other control systems and fault tolerant operation. Developments in semiconductor technologies have enabled the combination of logic, control and power switching devices on a single chip as shown in Fig. 1.45. Such smart power devices have the ability to monitor and control their own operation and to provide status reports on system behaviour, increasing the flexibility of the overall system. Fig. 1.46

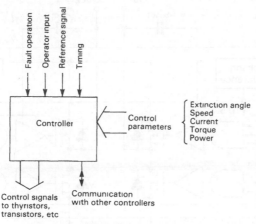

Fig. 1.44 Control system layout.

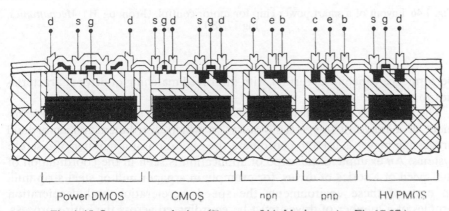

Fig. 1.45 Smart power device. (From pg 311 *Mechatronics*, Fig 17.27.)

shows the layout of a typical smart power chip for motor control, incorporating acceleration control and protection.

Other developments in smart power device technology include the use of thick film hybrid circuits and the use of advanced semiconductor materials such as silicon carbide and diamond films.

Electronics and power electronics

Electronics already plays a significant role in the real-time control of power electronic systems from variable speed drives to power transmission, enabling an optimization of system operation leading to increased performance capabilities and improved efficiencies. While microprocessor- and microcontroller-based systems provide the basis for many of these systems there is an increasing move towards the development of applications specific integrated circuits (ASICs) for specific functional requirements. The integration of the chip design process with the ability to incorporate a wide range of control laws within that process is likely to result in an ability to increase

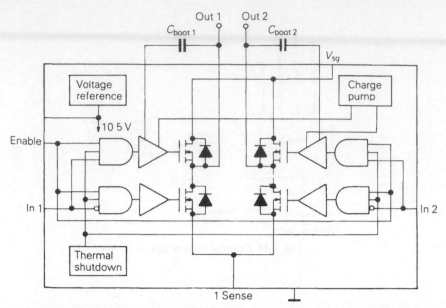

Fig. 1.46 Layout of a smart power chip for motor control. (From pg 311 *Mechatronics*, Fig 17.28.)

performance capabilities further with improved matching of drive characteristics to the load.

The increasing availability of processing power has also led to the development of a range of integrated systems in which distributed and embedded processing power is used to coordinate the operation of a number of discrete systems. An example of this type of application is seen in the coordination of the speed of a series of drives, for example in a paper mill or steel strip mill. In each of these environments the speed, acceleration and deceleration profiles of a number of drives must be coordinated across the entire process. While this could be achieved by means of one motor driving a single shaft with individual gearboxes used to select the individual speeds, this was inefficient, had mechanical problems such as shaft wind-up and a major maintenance overhead. By using individual microprocessor- or ASIC-based controllers with each machine operating under the control of a central control processor as suggested by Fig. 1.47, the need for the gearboxes is removed, resulting in a more compact, efficient and reliable installation.

Future developments are likely to include the incorporation of advanced control strategies such as fuzzy logic and neural networks at both the device and system level. Such controllers are likely to enable increased system functionally, perhaps including a self-learning and self-tuning capability to adjust the controller automatically to different machine characteristics.

The availability of power semiconductor devices with a wide range of current and voltage ratings and with a variety of switching performances has resulted in an increasing number of applications, including:

AC and DC motor drives
Servocontrol

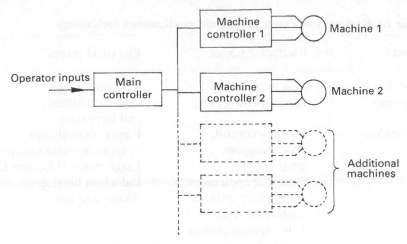

Fig. 1.47 Machine control system.

Voltage regulation
High voltage DC transmission (HVDC)
Switched mode power supplies
Uninterruptable power supplies
Heating controls
Lighting controls

Additionally, smart power devices are increasingly finding applications in a range of industries such as:

Aerospace
Automotive
Computers
Telecommunications
Consumer goods

Tables 1.1 and 1.2 (p. 36) together with Fig. 1.48 and 1.49 (p. 37) provide in simplified form a summary of the various power semiconductor types, their applications and likely market areas.

Systems integration

In the majority of applications, power semiconductor devices are associated with a control function exercised by means of the variation of the switching sequences of the devices themselves. In many instances, the application of microprocessor and microelectronics technologies as well as advanced fabrication techniques has resulted in the expansion of the control function leading to enhanced performance of the controlled element, for instance in the shaping of motor shaft characteristics, AC and DC servomotor control and pulse-width-modulated inverter drives.

Table 1.1 Summary of Power Electronic applications technology

Aspect	Electronic power	Electrical power
Power level	Low to medium	Medium to high
Frequency	High and increasing	Low to medium and increasing
Applications	Lighting controls	Power transmission
	Power supplies	Traction systems (AC or DC)
	Small motors	Large motors (AC and DC)
	Domestic appliances	Induction heating
	Automotive systems	Power supplies
	Audio and visual	
	Telecommunications	
	Aerospace	
Devices	Thyristors	Thyristors
	GTO thyristors	GTO thyristors
	BJTs	BJTs
	IGBTs	IGBTs
	Power MOSFETs	
	Modules	
	Smart power	
Construction	Flat pack	Flat pack
	Modules	Stud-base
	Surface mount	Capsule (pucks)

Table 1.2 Power semiconductor performance comparisons

Performance parameter	Thyristor	GTO thyristor	BJT	Power MOSFET	IGBT
Switching speed	**	****	****	*****	*****
Switching loss	**	**	***	*****	*****
On-state loss	**	**	****	**	***
Ease of turn-on	*****	***	***	****	****
Ease of turn-off	*	**	**	****	****
Current rating	*****	****	***	**	**
Voltage rating	*****	***	****	***	***
Surge current	*****	****	***	***	***

***** Best; * Worst.

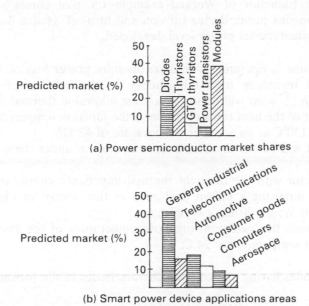

(a) Power semiconductor market shares

(b) Smart power device applications areas

Fig. 1.48 Predicted markets for the various types of power semiconductor devices.

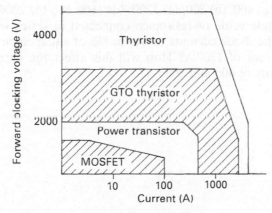

Fig. 1.49 Device ratings.

Problems

1.1 An approximation to the forward characteristic of a thyristor is as shown in the accompanying figure. Estimate the mean power loss in the thyristor for the following conditions:

(a) a constant current of 52 A for one-half cycle;
(b) a constant current of 22 A for one-third of a cycle;
(c) a constant current of 44 A for two-thirds of a cycle.

What will be the RMS current rating for each of the above load conditions?

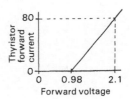

1.2 For the transistor of Worked example 1.5, plot curves showing the instantaneous power during turn-on and turn-off. Hence find the maximum instantaneous power level developed.

1.3 A thyristor is operating with a steady-state power loss of 32 W. If the thermal resistance from the junction to the heat transfer system is $0.81°C\,W^{-1}$, what will be the maximum allowable thermal resistance to ambient of the heat transfer system if the junction temperature is not to exceed 130°C at an ambient temperature of 42°C?

What will be the thyristor base temperature under these conditions?

1.4 A thyristor with the transient thermal impedance characteristic of Fig. 1.37 is supplying a pulsed load such as that shown in Fig. 1.38, with $P_{max} = 48$ W; $t = 20$ ms; $T = 100$ ms.

Estimate the maximum junction temperature of the thyristor if the ambient temperature is 24°C.

1.5 Two diodes having approximate characteristics in the forward direction:

Diode 1 $v = 0.88 + 2.44 \times 10^{-4} i$
Diode 2 $v = 0.96 + 2.32 \times 10^{-4} i$

are connected in parallel. Find the current in each diode if the total current is (a) 400, (b) 800, (c) 1200, (d) 1600 and (e) 2000 A.

What single value of resistance connected in series with each diode will bring the diode currents to within 8% of equal current sharing with a total current of 1200 A? How will this affect the current sharing at total currents of 400 and 2000 A?

2
Converters

- To understand what is meant by operation of a converter in the rectifying and inverting modes.
- To examine the operation of naturally commutated converters.
- To consider the operation of uncontrolled, fully-controlled and half-controlled converters.
- To develop the general equations describing converter behaviour.
- To understand the effect of firing delay and extinction angles on converter performance.
- To examine the effect of source inductance on commutation and to define overlap.
- To consider the function and operation of the freewheeling diode.
- To define power factor in relation to a converter circuit.

Naturally commutated converters

The function of a converter is that of transferring energy from an alternating current (AC) system to a direct current (DC) system (**rectification**) and, in certain cases, from a DC system to an AC system (**inversion**). In a naturally commutated converter using a combination of thyristors and diodes the turn-off of the conducting device takes place as a result of the action of the AC voltage following firing of the next device in sequence. No external circuitry is therefore required to turn the conducting device off. Three types of naturally commutated converter are available based on various combinations of thyristors and diodes; these are:

Uncontrolled converter or rectifier. An uncontrolled converter or rectifier is based on the use of diodes only and the output voltage is therefore determined solely by the magnitude of the AC supply. Energy can only be transferred from the AC system to the DC system.

Half-controlled converter. A half-controlled converter uses a combination of thyristors and diodes enabling control of the DC output voltage to be achieved by varying the firing angle of the thyristors. Energy can only be transferred from the AC system to the DC system. The half-controlled converter is cheaper than a fully-controlled converter of similar rating.

Fully-controlled converter. A fully-controlled converter uses only thyristors, with control of the DC output voltage determined by the thyristor firing

Rectification is the process of converting a bidirectional (alternating) current or voltage into a unidirectional (direct) current or voltage. Power flow in rectifying mode is from the AC system to the DC system.

Inversion is the process of transferring energy from a DC system to an AC system. Power flow in inverting mode is from the DC system to the AC system.

39

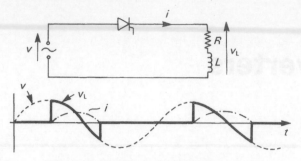

Fig. 2.1a Thyristor with inductive load.

angles. Operation can either be in rectifying mode with energy transferred from the AC system to the DC system or in inverting mode with energy transferred from the DC system to the AC system.

Pulse number

Converters are often described by their pulse number. This is the number of discrete switching operations involving load transfer (commutation) between individual diodes or thyristors during the period covered by one cycle of the AC voltage waveform. The pulse number is therefore directly related to the repetition period of the DC voltage waveform.

Commutating or freewheeling diode

It was shown in Chapter 1 that if a thyristor or diode is used to supply an inductive load, then the load voltage will reverse towards the end of the conduction interval as shown in Fig. 2.1(a). Particularly with uncontrolled or half-controlled converters, a commutating or freewheeling diode is used as in Fig. 2.1(b) to transfer current away from the main power devices, eliminating the voltage reversal. In this way, the commutating diode facilitates the recovery of the main power devices into their blocking state ready for the next switching cycle.

Current through the commutating diode is maintained during the off period of the main power devices by the energy stored in the magnetic field of the load inductance.

Converter operation

Figure 2.2(a) shows a full-wave rectifier using diodes, while Fig. 2.2(b) is essentially the same circuit but with the diodes replaced by thyristors. In the case of the circuit of Fig. 2.2(a), transfer of load current from diode to diode occurs automatically as a result of the action of the AC supply. Each diode conducts for one half cycle of the AC waveform in turn drawing a non-sinusoidal alternating current from the supply.

In the case of the circuit of Fig. 2.2(b), the point-on-wave at which the thyristors are fired is defined by the firing angle or firing delay angle α. The firing angle is measured from the earliest point at which the thyristor could

The process of transferring load from one thyristor to another is referred to as **commutation**

Also referred to as a biphase, half-wave rectifier.

This definition of firing angle is applied to all the converters considered. The reponse of the corresponding diode circuit, thyristors replaced by diodes, can therefore be obtained by putting $\alpha = 0°$.

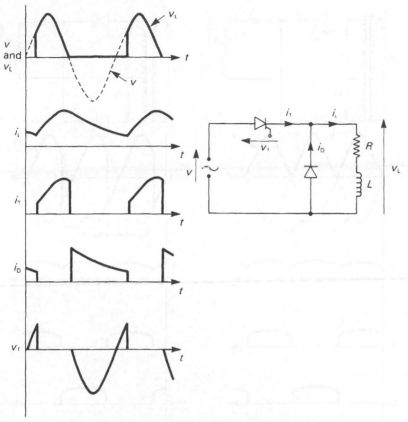

Fig. 2.1b Effect of commutating diode.

begin to conduct which is also the point at which a diode replacing the thyristor would begin to conduct.

For the circuit shown with an inductive load such that the load current is continuous throughout, then at the instant of firing thyristor T_2, with thyristor T_1 conducting, the voltage v_2 is more positive than the voltage v_1, resulting in a transfer of the load current from thyristor T_1 to thyristor T_2.

This is typical of many industrial loads such as motors.

The mean load voltage is obtained by averaging the converter output voltage over the conduction interval. For the circuit of Fig. 2.2(a),

$$V_{\text{mean}} = \frac{1}{\pi} \int_0^{\pi} \hat{V} \sin\omega t \, \mathrm{d}(\omega t) = \frac{2\hat{V}}{\pi} \qquad (2.1)$$

$\hat{V}$ is the peak amplitude of the supply voltage.

For that of Fig. 2.2(b):

$$V_{\text{mean}} = \frac{1}{\pi} \int_{\alpha}^{\alpha + \pi} \hat{V} \sin\omega t \, \mathrm{d}(\omega t) = \frac{2\hat{V} \cos\alpha}{\pi} \qquad (2.2)$$

Put $\alpha = 0°$ in Equation 2.2 and compare with the result of Equation 2.1.

Figure 2.3 shows the circuits and waveforms associated with the single-phase fully-controlled and half-controlled bridges. The conduction period for the fully-controlled bridge is π and that of the half-controlled bridge is $\pi - \alpha$ as a result of the action of the commutating diode. The effect of the commutating diode on the operation of the half-controlled bridge is clearly seen from

For a practical (non-ideal) thyristor or diode the mean voltage will be reduced by the device forward voltage drop.

41

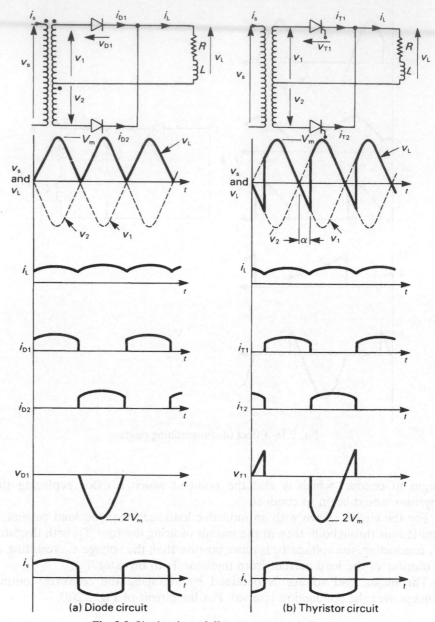

Fig. 2.2 Single-phase full-wave converter circuits.

the output voltage and current waveforms. The respective load voltages, ignoring diode and thyristor forward voltage drops are:

Fully-controlled bridge

$$V_{\text{mean}} = \frac{1}{\pi} \int_{\alpha}^{\alpha + \pi} \hat{V} \sin\omega t \; \mathrm{d}(\omega t) = \frac{2\hat{V} \cos\alpha}{\pi} \tag{2.3}$$

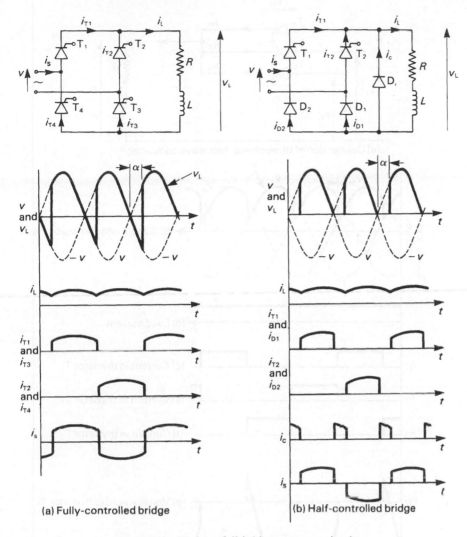

(a) Fully-controlled bridge

(b) Half-controlled bridge

Fig. 2.3 Single-phase full-bridge converter circuits.

Half-controlled bridge

$$V_{\mathrm{mean}} = \frac{1}{\pi} \int_{\alpha}^{\pi} \hat{V} \sin\omega t \ \mathrm{d}(\omega t) = \frac{2\hat{V}(1 + \cos\alpha)}{\pi} \tag{2.4}$$

In the case of Equation 2.3 for values of $\alpha > \pi/2\,(90°)$, V_{mean} becomes negative. As the direction of current through the thyristors cannot reverse, the effect is to transfer power from the DC system to the AC system in which case the converter is operating in inverting mode. This should be compared with Equation 2.4 which will be positive for all values of α.

See page 50 for discussion of inversion and inverting mode.

Single-phase converters are limited to powers of a few kilowatts. Where higher power levels are required, converters based on three-phase systems are available. The three-phase, half-wave converter circuit of Fig. 2.4(a) forms

Details of three-phase systems are given in Appendix B.

43

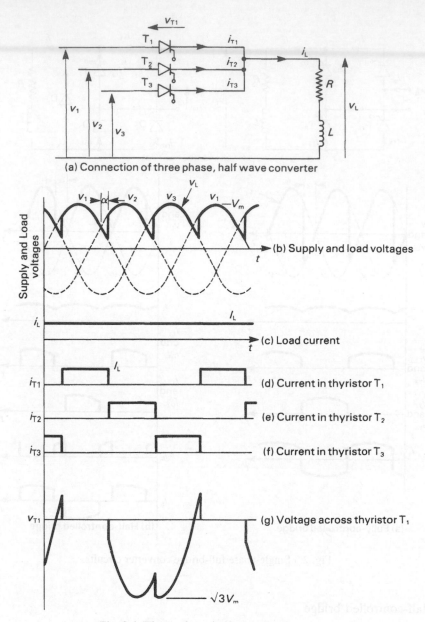

Fig. 2.4 Three-phase half-wave converter.

For the three-phase,
half-wave converter the
voltage crossover occurs at
$\pi/6$ (30°) after the
corresponding voltage zero.

the basis of many of the more complex circuits and will therefore be used for the initial discussion of converter operation.

Assuming ideal thyristors together with an inductive load drawing a continuous, constant current, the load voltage waveform for the three-phase, half-wave converter at a firing angle less than $\pi/2$ will have the general form shown in Fig. 2.4(b). The firing angle α is measured from the crossover points of the voltage waveform as these represent the instant at which the diodes in the equivalent diode circuit would begin to conduct.

The resulting conduction period for each thyristor is then one third of a cycle ($2\pi/3$ or $120°$). The mean voltage is calculated as before by averaging the output voltage over the repetition period of the output voltage waveform, which in this case corresponds to the conduction period of an individual thyristor:

$$V_{\text{mean}} = \frac{3}{2\pi} \int_{\frac{\pi}{6} + \alpha}^{\frac{5\pi}{6} + \alpha} \hat{V} \sin\omega t \, \mathrm{d}(\omega t) = \frac{3\sqrt{3}}{2\pi} \hat{V} \cos\alpha \qquad (2.5)$$

$\hat{V}$ in Equation 2.5 is the amplitude of the phase voltage of the three-phase supply.

The RMS current in each thyristor is obtained by integrating over one cycle of the supply when, for a constant load current I_L:

$$I_{\text{RMS}} = \left[\frac{1}{2\pi} \int_{\alpha}^{\alpha + \frac{2\pi}{3}} I_L^2 \, \mathrm{d}(\omega t) \right]^{\frac{1}{2}} = \frac{I_L}{\sqrt{3}} \qquad (2.6)$$

So far, the three-phase, half-wave converter has been shown connected directly to the three-phase supply. In practice, the converter will often be supplied by means of a transformer. If a simple star–star transformer such as that shown in Fig. 2.5 is used, then a unidirectional current will flow in each phase and may well result in the DC magnetization of the transformer core. In order to prevent this occurring, the interconnected star secondary winding of Fig. 2.6(a) is used, ensuring a bidirectional current in each of the transformer primary windings as in Fig. 2.6(b).

The DOT convention is used with mutually coupled coils to relate the sense of the induced voltage to the direction of the source current. For a current direction as shown below, the voltage in the coupled coil has the polarity indicated.

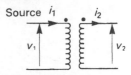

Also referred to as a star-zigzag winding.

A three-phase, half-wave, fully-controlled converter is connected to a 380 V (line) supply. The load current is constant at 32 A and is independent of firing angle.

Worked example 2.1

Find the mean load voltage at firing angles of $0°$ and $45°$, assuming that the thyristors have a forward voltage drop of 1.2 V. What current and peak reverse voltage rating will the thyristors require and what will be the average power dissipation in each thyristor?

The effect of the thyristor forward voltage drop will be to reduce the mean load voltage from theoretical value. Hence

$$\hat{V} = \frac{380 \times \sqrt{2}}{\sqrt{3}} = 310.3 \text{ V}$$

Phase voltage=(Line voltage) $/\sqrt{3}$ for a three-phase system (Appendix 1)

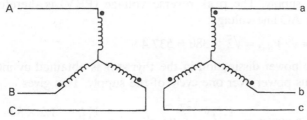

Fig. 2.5 Star–star transformer.

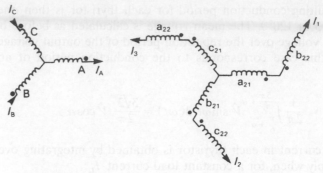

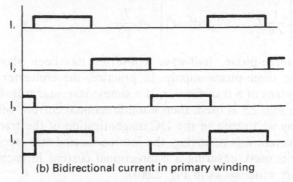

(a) Windings a_{21} and a_{22} couple with winding A etc

(b) Bidirectional current in primary winding

Fig. 2.6 Interconnected star winding.

V_t is the thyristor forward voltage drop.

$$V_{mean} = \frac{3\sqrt{3}}{2\pi} \hat{V} \cos\alpha - V_t = 256.6 \cos\alpha - 1.2$$

$\alpha = 0°$

$$V_{mean,\,0°} = 256.6 \cos0° - 1.2 = 255.4\,V$$

$\alpha = 45°$

$$V_{mean,45°} = 256.6 \cos45° - 1.2 = 180.2\,V$$

Ratings

From Equation 2.6,

$$I_{RMS} = \frac{32}{\sqrt{3}} = 18.47\,A$$

From Fig. 2.4 the reverse voltage across the thyristor can be seen to be the difference between the two phase voltages, i.e. the line voltage of the three-phase supply. The peak reverse voltage (PRV) is therefore the peak value of the AC line voltage:

$$PRV = \sqrt{2}\,V_{line} = \sqrt{2} \times 380 = 537.4\,V$$

The average power dissipation in the thyristor is obtained by integrating the instantaneous power over one cycle of the supply. This gives

$$\text{Average power} = \frac{1}{2\pi} \int_{\alpha}^{\alpha + \frac{2\pi}{3}} v_t i_t \, d(\omega t) = \frac{V_t I_t}{3} = \frac{1.2 \times 32}{3} = 12.8\,W$$

Overlap

So far no account has been taken of the effect of source inductance on converter behaviour. Consider the three-phase, half-wave bridge at the instant of firing thyristor T_2. With a constant load current the effect of firing thyristor T_2 is to set up a circulating current i between thyristors T_1 and T_2 as shown in Fig. 2.7. This circulating current will increase from zero at the instant of firing thyristor T_2 until it equals the load current I_L, at which point the current in thyristor T_1 becomes zero when this thyristor is turned off and commutation is completed. The interval during which both thyristors are conducting is referred to as the overlap period and is defined by the overlap angle γ. This angle may be calculated by reference to Fig. 2.7 when, ignoring thyristor voltage drops:

$$v_2 - v_1 = 2L \, di/dt \tag{2.7}$$

Putting $t = 0$ at the instant of firing thyristor T_2, then

$$v_2 - v_1 = v_{\text{line}} = \sqrt{3} \, \hat{V} \sin(\omega t + \alpha) \tag{2.8}$$

Combining Equations 2.6 and 2.7:

$$di = \frac{\sqrt{3}\hat{V}}{2L} \sin(\omega t + \alpha) dt$$

Integrating this equation from $t = 0$ to t:

$$i = \frac{\sqrt{3}\hat{V}}{2\omega L} \{\cos \alpha - \cos(\omega t + \alpha)\} \tag{2.9}$$

Commutation is complete when $i = I_L$, i.e. when $\omega t = \gamma$. Hence

$$I_L = \frac{\sqrt{3}\hat{V}}{2\omega L} [\cos \alpha - \cos(\gamma + \alpha)] \tag{2.10}$$

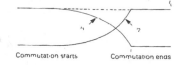

Thyristor currents during commutation.
Load current $= I_L = i_1 - i_2$

Commutation starts   Commutation ends

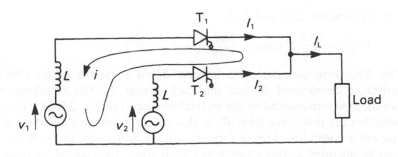

Fig. 2.7 Commutation between thyristor T_1 and T_2 including source inductance.

47

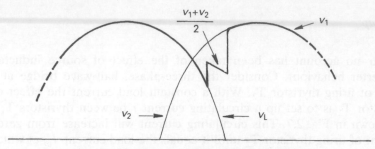

Fig. 2.8 Converter voltage during overlap.

Since both thyristors are conducting together. Assumes identical source inductances values.

$v_1 = \hat{V} \sin(\omega t + 2\pi/3)$

$v_2 = \hat{V} \sin \omega t$

During commutation the load voltage will be the mean of V_1 and V_2 as shown in Fig. 2.8, resulting in a change in the converter mean voltage at the load to

$$V_{\text{mean}} = \frac{3}{2\pi}\left\{ \int_{\alpha + \gamma + \frac{\pi}{6}}^{\alpha + \frac{5\pi}{6}} \hat{V} \sin(\omega t)\, d(\omega t) \right.$$

$$+ \frac{1}{2}\int_{\alpha + \frac{\pi}{6}}^{\alpha + \gamma + \frac{\pi}{6}} \hat{V}\left[\sin\left(\omega t + \frac{2\pi}{3}\right) + \sin \omega t\right] d(\omega t)\left.\right\}$$

$$= \frac{3\sqrt{3}\hat{V}}{4\pi}\{\cos\alpha + \cos(\alpha + \gamma)\} \tag{2.11}$$

Now let

$$V_{\text{mean}} = V_O - \Delta V_d$$

where V_O is the mean output voltage of the converter in the absence of overlap and ΔV_d is the change in converter output voltage as a result of overlap. From Equation 2.5,

$$V_O = \frac{3\sqrt{3}}{2\pi}\hat{V}\cos\alpha$$

Hence

$$\Delta V_d = \frac{3\sqrt{3}}{4\pi}\hat{V}\{\cos\alpha - \cos(\alpha + \gamma)\} \tag{2.12}$$

Combining Equations 2.10 and 2.12:

$$\Delta V_d = \frac{3\omega L}{2\pi}I_L \tag{2.13}$$

From Equations 2.11 and 2.13

$$V_{\text{mean}} = \frac{3\sqrt{3}}{2\pi}\hat{V}\cos\alpha - \frac{3\omega L}{2\pi}I_L = V_O - R_r I_L \tag{2.14}$$

The ΔV_d term can therefore be considered in terms of an effective DC resistance component R_r and the load current I_L. The rectifying converter can then be represented by the equivalent circuit of Fig. 2.9. It is important to note that the resistance term R_r in this circuit represents a voltage drop **only**, and **not** a power loss. Device forward voltage drops and lead resistances could also be included in this equivalent circuit when they would represent **both** a voltage drop **and** a power loss.

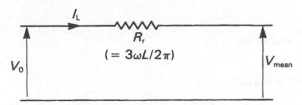

$$V_o \xrightarrow{\quad I_L \quad} \substack{R_r \\ (= 3\omega L/2\pi)} \longrightarrow V_{mean}$$

Fig. 2.9 Simple equivalent circuit for three-phase half-wave converter in rectifying mode.

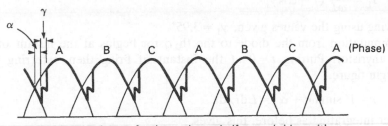

(a) Load voltage of a three-phase, half-wave bridge with continuous load current and overlap

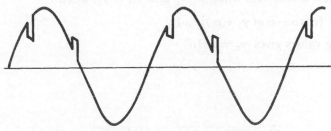

(b) Phase A, voltage at thyristor anode

Fig. 2.10 Operation of a three-phase half-wave bridge with overlap.

Figure 2.10 shows the effect of overlap on the output voltage, phase currents and source voltage of a three-phase, half-wave converter.

A single-phase, full-wave converter as shown in the margin figure is supplied from a 120 V, 50 Hz supply with a source inductance of 0.32 mH. Assuming the load current is continuous at 4 A, find the overlap angles for (i) transfer of current from a conducting thyristor to the commutating diode and (ii) from the commutating diode to a thyristor when the firing angle is 15°.

Commutation from thyristor to diode begins at time $t = 0$, the instant when the load voltage starts to reverse (ignoring device voltage drops). Hence at the onset of commutation and referring to the margin figure,

$$\nu = -\hat{V}\sin\omega t = -L\,\mathrm{d}i/\mathrm{d}t$$

From this equation

$$\mathrm{d}i = \hat{V}\sin(\omega t)\mathrm{d}t/L$$

Worked example 2.2

Single-phase, full-wave converter with commutating diode.

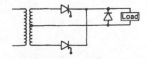

Commutation from thyristor diode.

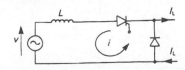

Integrating

$$i = \frac{\hat{V}}{L} \int_0^t \sin(\omega t) \mathrm{d}t$$

$$= \frac{\hat{V}}{\omega L}(1 - \cos \omega t)$$

Commutation is complete when $i = I_L$, at which point $\omega t = \gamma_1$.
Thus

$$I_L = \frac{\hat{V}}{\omega L}(1 - \cos \gamma_1)$$

Solving using the values given, $\gamma_1 = 3.95°$.

Commutation from the diode to the thyristor begins at the instant of firing the thyristor. Putting $t = 0$ at the instant of firing then, referring to the margin figure:

$$v = \hat{V} \sin(\omega t + \alpha) = L \mathrm{d}i/\mathrm{d}t$$

After integrating as before, this gives

$$i + \hat{V}[\cos \alpha - \cos(\omega t - \alpha)]/\omega L$$

Commutation is complete when $i = I_L$ and $\omega t = \gamma_2$, when

$$I_L = \hat{V}[\cos \alpha - \cos(\gamma_2 + \alpha)]/\omega L$$

Substituting values gives $\gamma_2 = 0.516°$

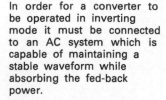

Inversion

In order for a converter to be operated in inverting mode it must be connected to an AC system which is capable of maintaining a stable waveform while absorbing the fed-back power.

Figure 2.11 shows the variation of mean load voltage with firing angle for a three-phase, half-wave converter in the absence of overlap. For firing angles greater than 90°, the converter mean voltage becomes negative and, since the direction of current through the thyristor cannot reverse, the direction of power flow is now from the DC side of the converter back into the AC supply. Under these conditions the converter is operating in inverting mode. Figure 2.12 shows the load voltage waveforms for the three-phase, half-wave converter at various firing angles.

The firing advance angle β is measured from the latest point at which ideal, instantaneous commutation could occur in the absence of overlap.

When operating in inverting mode, the point-on-wave at which a thyristor is fired is more usually defined by reference to the **firing advance angle** β

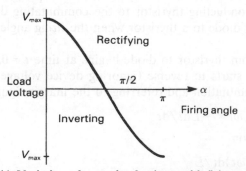

Fig. 2.11 Variation of mean load voltage with firing angle.

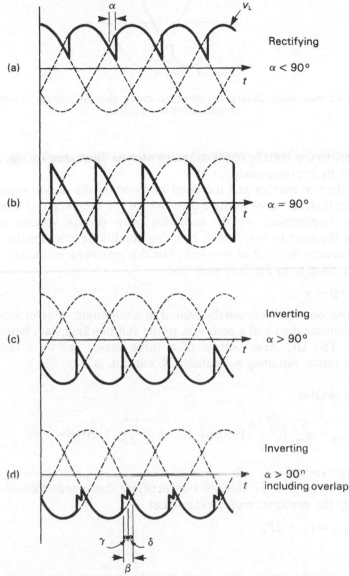

(a) Rectifying α < 90°

(b) α = 90°

(c) Inverting α > 90°

(d) Inverting α > 90° including overlap

Fig. 2.12 Three-phase half-wave converter showing effect of firing angle variation.

rather than the firing delay angle α. The relationship between α and β is given by

$$\beta = 180° - \alpha \qquad (2.15)$$

and applies to converters of any pulse number.

Commutation between any pair of thyristors will only occur if the anode voltage of the thyristor undergoing turn-on is greater (more positive) than the anode voltage on the thyristor being turned off throughout the whole of the period of overlap. If this is not the case and the anode voltages, though initially of the correct relationship, first become equal and then reverse before the current transfer is complete, then conduction, and hence current,

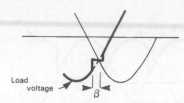

Fig. 2.13 Commutation failure in inverting mode due to voltage reversal before commutation is complete.

will revert to the initially conducting thyristor as illustrated by Fig. 2.13 for a converter in inverting mode.

The effect of overlap and the need to maintain the anode voltage of the oncoming thyristor more positive than that of the outgoing thyristor together with the requirement for the provision of a reverse voltage across the outgoing thyristor to re-establish blocking conditions modifies the operating conditions near the limit of $\alpha = 180°$. For this reason an extinction angle δ is often specified as in Fig. 2.14 such that

The extinction angle δ is also referred to as the recovery angle.

$$\delta = \beta - \gamma \tag{2.16}$$

The extinction angle δ is usually limited to a minimum value of around 5° to ensure commutation and a reversion to the full blocking state before voltage reversal. The DC mean voltage of a three-phase, half-wave converter in inverting mode, assuming a constant DC current, is

Ignoring overlap

$$V_{\text{mean}} = \frac{3}{2\pi} \int_{\frac{\pi}{6} - \beta}^{\frac{5\pi}{6} - \beta} \hat{V} \sin \omega t \, \mathrm{d}(\omega t) = \frac{3\sqrt{3}}{2\pi} \hat{V} \cos \beta = V_O \tag{2.17}$$

Including overlap

As is the case with the converter in rectifying mode, the effect of overlap is to modify the converter mean voltage. Let

$$V_{\text{mean}} = V_O + \Delta V_d \tag{2.18}$$

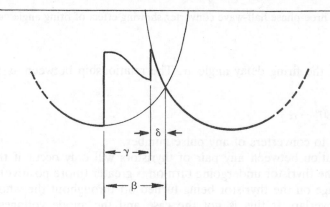

Fig. 2.14 Extinction angle.

when

$$\Delta V_d = \frac{3\sqrt{3}}{4\pi} \hat{V}\{\cos\beta - \cos(\beta - \gamma)\}$$

Equation 2.10 can be rewritten for inverting mode as

$$I_L = -\frac{\sqrt{3}}{2\omega L} \hat{V}\{\cos\beta - \cos(\beta - \gamma)\} \tag{2.19}$$

In which case, combining equations 2.17, 2.18 and 2.19

$$V_{mean} = \frac{\sqrt{3}}{2\omega L} \hat{V}\cos\beta + \frac{3\omega L}{2\pi} I_L = V_O + R_i I_L \tag{2.20}$$

where $\Delta V_d = R_i I_L$ and $R_i = 3\omega L/2\pi$.

Equation 2.20 can be represented by the inverter equivalent circuit of Fig. 2.15 with R_i representing a voltage drop **only** and **not** a power loss.

A three-phase, half-wave converter is operating in the inverting mode connected to a 415 V (line) supply. If the extinction angle is 18° and the overlap 3.8° find the mean voltage at the load.

Worked example 2.3

Using Equation 2.17,

$$V_{mean} = \frac{3\sqrt{3}}{4\pi} \times \frac{415\sqrt{2}}{\sqrt{3}} \times (\cos 18° + \cos 14.2°) = 2.69.1 \text{ V}$$

Converter equations

The converter equations developed so far have been for the three-pulse, three-phase, half-wave converter. Similar analyses can be used to develop the equations for a general p-pulse, fully-controlled converter. Referring to Fig. 2.16:

Rectifying
Ignoring overlap

$$V_O = \frac{p}{2\pi} \int_{\alpha - \frac{\pi}{p}}^{\alpha + \frac{\pi}{p}} \hat{V} \cos(\omega t)\, d(\omega t) = \frac{p}{\pi} \hat{V} \sin(\pi/p) \cos\alpha \tag{2.21}$$

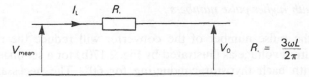

Fig. 2.15 Equivalent circuit for three-phase half-wave converter in inverting mode.

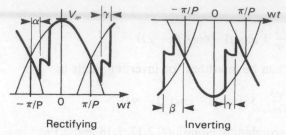

Fig. 2.16 Output voltage waveforms for a *p*-pulse converter.

Including overlap

$$V_{\text{mean}} = \frac{p}{2\pi} \left\{ \int_{\alpha+\gamma-\frac{\pi}{p}}^{\alpha+\frac{\pi}{p}} \hat{V} \cos(\omega t)\, \mathrm{d}(\omega t) \right.$$

$$+ \frac{1}{2} \int_{\alpha-\frac{\pi}{p}}^{\alpha+\gamma-\frac{\pi}{p}} \hat{V} \left[\cos(\omega t) + \cos\left(\omega t + \frac{2\pi}{p}\right) \right] \mathrm{d}(\omega t) \right\}$$

$$= \frac{p}{2\pi} \hat{V} \sin\left(\frac{\pi}{p}\right) [\cos\alpha + \cos(\alpha+\gamma)]$$

$$= \frac{p}{\pi} \hat{V} \sin\left(\frac{\pi}{p}\right) \cos\alpha - \frac{p\omega L}{2\pi} I_{\text{L}} \qquad (2.22)$$

Inverting
Ignoring overlap

$$V_{\text{O}} = \frac{p}{2\pi} \int_{-\beta-\frac{\pi}{p}}^{-\beta+\frac{\pi}{p}} \hat{V} \cos(\omega t)\, \mathrm{d}(\omega t) = \frac{p}{\pi} V_{\text{m}} \sin(\pi/p) \cos\beta \qquad (2.23)$$

Including overlap

$$V_{\text{mean}} = \frac{p}{2\pi} \left\{ \int_{-\beta-\frac{\pi}{p}+\gamma}^{-\beta+\frac{\pi}{p}} \hat{V} \cos(\omega t)\, \mathrm{d}(\omega t) \right.$$

$$+ \frac{1}{2} \int_{-\beta-\frac{\pi}{p}}^{-\beta-\frac{\pi}{p}+\gamma} \hat{V} \left[\cos\omega t + \cos\left(\omega t + \frac{2\pi}{p}\right) \right] \mathrm{d}(\omega t) \right\}$$

$$= \frac{p}{2\pi} \hat{V} \sin\left(\frac{\pi}{p}\right) [\cos\beta + \cos(\beta+\gamma)]$$

$$= \frac{p}{\pi} \hat{V} \sin\left(\frac{\pi}{p}\right) \cos\beta + \frac{p\omega L}{2\pi} I_{\text{L}} \qquad (2.24)$$

Converters with higher pulse numbers

In Fig. 2.17 and subsequently, overlap is ignored for clarity and simplicity. It is present in any real circuit.

Increasing the pulse number of the converter will reduce the ripple in the converter output voltage as illustrated by Fig. 2.17(b) for a six-phase, half-wave converter with each thyristor conducting for 60°. The necessary six-phase supply could be obtained from a three-phase supply by using the transformer

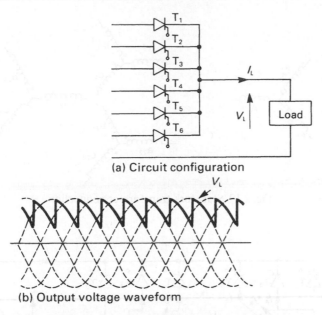

(a) Circuit configuration

(b) Output voltage waveform

Fig. 2.17 Six-phase half-wave fully-controlled converter.

connection of Figure 2.18. However, this simple connection in which current flows in each leg of the primary winding for only one-third of a cycle, is not normally used as it introduces high levels of harmonic currents into the primary system. To reduce the levels of primary harmonic current, the star–fork connection of Fig. 2.19 could be used in which case current flows in each phase of the primary winding for two-thirds of a cycle.

Such unconventional transformer windings are expensive and the dual converter arrangement of Fig. 2.20(a) may be used instead in which the star points of the two secondary windings of the supply transformer are connected by an interphase reactor. Each group of thyristors (T_1, T_3, T_5 and T_2, T_4, T_6) then operates as a conventional three-phase, half-wave bridge with each thyristor conducting for 120°. The load voltage under these conditions is the mean of the output voltages of the individual three-phase, half-wave groups as shown in Fig. 2.20(b). In order to balance the voltage distribution around the circuit, potential difference across the interphase reactor is then the difference between the output voltages of the individual three-phase, half-wave groups.

The operation of the interphase reactor depends on the presence of a circulating magnetizing current which flows between the two star points with

Third-harmonic current in the circuit of Fig. 2.18 has a magnitude of $4I_L/3\pi$. Third-harmonic current in the circuit of Fig. 2.19 is zero. Assumes no overlap and a constant load current.

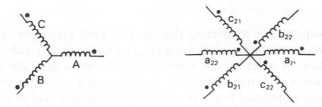

Fig. 2.18 Transformer configuration giving an effective six-phase output.

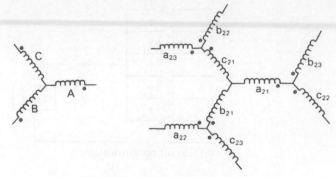

Fig. 2.19 Star–fork transformer connection.

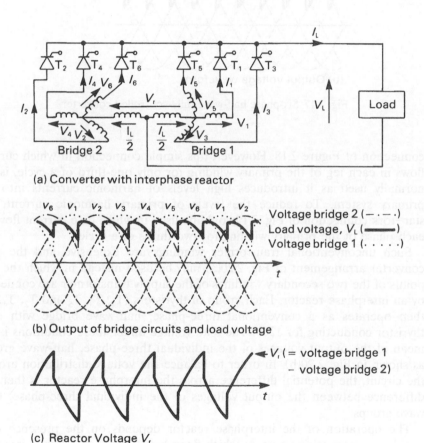

(a) Converter with interphase reactor

Voltage bridge 2 ($- \cdot - \cdot$)
Load voltage, V_L (———)
Voltage bridge 1 ($\cdots$)

(b) Output of bridge circuits and load voltage

V_r (= voltage bridge 1 − voltage bridge 2)

(c) Reactor Voltage V_r

Fig. 2.20 Converter with interphase reactor.

a return path via the conducting thyristors of each group, in one case as a reverse current, causing a slight imbalance between the currents in each half of the reactor. In order for the magnetizing current to flow, the load current must be greater than the magnetizing current and a converter of this type is therefore often operated into a permanently connected load of sufficient size to ensure correct operation for all added loads.

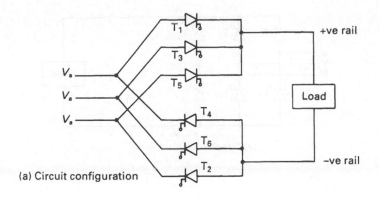

(a) Circuit configuration

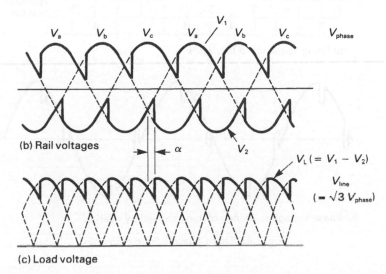

(b) Rail voltages

(c) Load voltage

Fig. 2.21 Three-phase full-wave converter formed by connection of two three-phase half-wave converters.

The connections of both the six-phase, half-wave converter and the dual converter with interphase reactor are both relatively inefficient in terms of their need for special transformers and additional inductive reactors.

Figure 2.21(a) shows two three-phase, half-wave converters connected to operate off the positive and negative half-cycles of the supply waveform respectively. The resulting positive and negative rail voltages are then as shown in Fig. 2.21(b). The voltage across the load is the difference between these rail voltages and has the form shown in Fig. 2.21(c).

Compare the waveform of Fig. 2.21(b) with that for the six-phase, half-bridge of Fig. 2.17(b).

The circuit of Fig. 2.21(a) can be redrawn in the form of Fig. 2.22 which is the familiar three-phase, six-pulse, full-bridge converter in which each thyristor conducts for 120° with a commutation taking place as shown by Fig. 2.22(b).

Since in normal operation, two thyristors are conducting together, an appropriate pair of thyristors must be gated together to initiate the operation of the converter. In practice, this means that one thyristor is always supplied with two firing signals 60° apart, the second signal having no effect on the

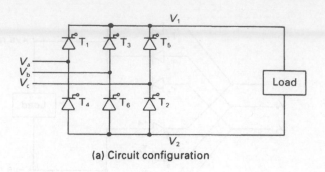

(a) Circuit configuration

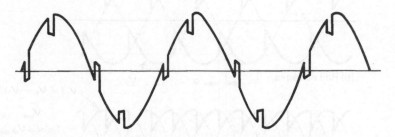

(b) Firing sequence of thryistors

(c) Phase voltage showing notching caused by overlap

Fig. 2.22 Three-phase bridge converter.

receiving thyristor once operation has been fully initiated and the conducting sequence established.

For applications such as high-voltage DC (HVDC) transmission, higher pulse number circuits may be required. Figure 2.23 shows two ways in which two six-pulse bridges may be combined using transformers with a 30° phase shift between their respective pairs of secondary voltages in order to produce effective 12-pulse operation as seen from the AC supply. Higher effective pulse numbers can similarly be obtained by combining the basic six-pulse full-bridge with transformers of varying secondary phase shifts.

High-voltage DC (HVDC) transmission is discussed in Chapter 6.

Worked example 2.4

A DC link consists of two six-pulse, fully-controlled bridge converters of the type shown in Fig. 2.20(a) connected by a transmission line of 0.2 Ω resistance and used to connect a three-phase, 415 V (line), 50 Hz system to a three-phase, 380 V (line), 60 Hz system. The source inductance of the 50 Hz system is 1 mH per phase and that of the 60 Hz system 1.25 mH per phase.

If the DC link is carrying a constant DC current of 50 A and delivering 15 kW into the 60 Hz system, find the firing advance angle of the inverter and the firing angle of the rectifier.

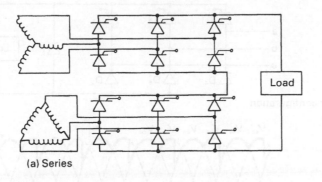

(a) Series

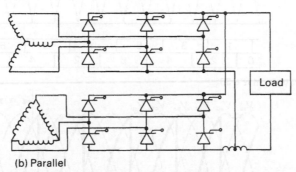

(b) Parallel

Fig. 2.23 Twelve-pulse bridge configurations using two six-phase bridges.

The basic rectifier and inverter equivalent circuits of Figs. 2.9 and 2.15 can be combined with the DC link resistance to give the system equivalent circuit shown. The values of R_r and R_i can then be calculated from the given data:

Note frequencies used in calculating R_i and R_r.

$$R_r = \frac{6 \times 100 \times \pi \times 10^{-3}}{2\pi} = 0.3\,\Omega$$

$$R_i = \frac{6 \times 120 \times \pi \times 1.25 \times 10^{-3}}{2\pi} = 0.45\,\Omega$$

For the inverter the mean voltage can be found from the power flow and DC current:

$$V_{\mathrm{mean,i}} = \frac{1\,500}{50} = 300\,\mathrm{V}$$

The mean voltage in the absence of overlap can then be found:

$$V_{0,i} = V_{\mathrm{mean,i}} - R_i I_L = 300 - (50 \times 0.45) = 277.5\,\mathrm{V}$$

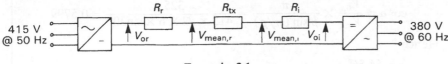

Example 2.1.

59

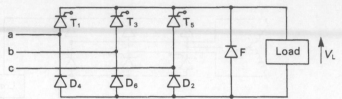

(a) Circuit configuration

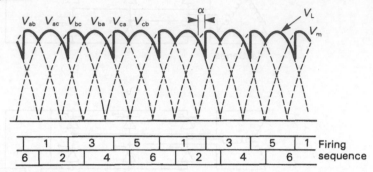

(b) Output waveform $\alpha < 60°$

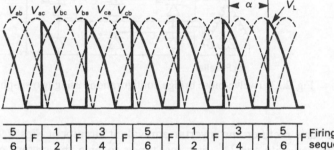

(c) Output waveform $\alpha > 60°$

Fig. 2.24 Half-controlled three-phase bridge converter.

Using Equation 2.20 with $p = 6$ and $\hat{V}$ the peak value of the line voltage:

$$\cos\beta = \frac{V_{0,i}}{V_{m,i}} \times \frac{\pi}{p\sin(\pi/p)} = \frac{277.5 \times \pi}{380 \times \sqrt{2} \times 6 \times \sin 30°} = 0.5408 \quad (2.25)$$

Hence

$$\beta = 48.53°$$

The rectifier mean voltage, including overlap, can now be found:

$$V_{mean,r} = V_{mean,i} + R_t I_L = 300 + (50 \times 0.2) = 310\,\text{V}$$

The mean voltage, ignoring overlap, can then be obtained using the load current and R_r:

$$V_{0,r} = V_{mean,r} + I_L R_r = 310 + (50 \times 0.3) = 325\,\text{V}$$

Using Equation (2.20) with $p = 6$:

$$\cos\alpha = \frac{V_{0,r}}{V_{m,r}} \times \frac{\pi}{p\sin(\pi/p)} = \frac{325\,\pi}{415 \times \sqrt{2} \times 6 \times \sin 30°} = 0.5799$$

Hence

$$\alpha = 54.56°$$

Half-controlled converters

With the exception of the single-phase, full-bridge form of Fig. 2.3(b), all the converters considered so far have been fully-controlled. Figure 2.24 shows the circuit configuration and output voltage waveforms for a three-phase, half-controlled, full-bridge converter incorporating a commutating diode.

Consider the operation of this converter when the firing angle is less than 60°. In this case, the output waveform will have the form shown in Fig. 2.24(b) with load being transferred from one thyristor to the next thyristor in sequence on the firing of the appropriate thyristor. Load is automatically transferred in sequence between the diodes as a result of the action of the AC supply voltage waveform. The commutating diode plays no part in the operation of the converter under these conditions.

The output voltage of the converter with $\alpha \leq 60°$ is then

$$V_{\text{mean}} = \frac{3}{2\pi}\left[\int_{\alpha+\frac{\pi}{3}}^{\frac{2\pi}{3}} \hat{V}\sin\omega t\, \mathrm{d}(\omega t) + \int_{\frac{\pi}{3}}^{\frac{2\pi}{3}+\alpha} \hat{V}\sin\omega t\, \mathrm{d}(\omega t)\right]$$

$$= \frac{3}{2\pi}\,\hat{V}\,(1+\cos\alpha) \qquad (2.26)$$

When the firing angle reaches 60°, by inspection of the circuit of Fig. 2.24 and the voltage waveforms of Fig. 2.24(b), it can be seen that the load voltage would attempt to become negative but would be prevented from doing so by the action of the conducting thyristor and its associated diode. Current would then continue to flow through the load via the appropriate thyristor/diode pair.

Thyristor T_1 is associated with diode D_4, etc.

The inclusion of the commutating diode causes the load current to be transferred from the main thyristor, enabling full blocking conditions to be restored before the firing of the next thyristor in sequence. In this case, the output voltage waveform and firing/conduction sequence is as shown in Fig. 2.24(c) when the mean load voltage becomes:

$$V_{\text{mean}} = \frac{3}{2\pi}\int_{\alpha}^{\pi} \hat{V}\sin\omega t\, \mathrm{d}(\omega t) = \frac{3}{2\pi}\,\hat{V}\,(1+\cos\alpha) \qquad (2.27)$$

which is the same as Equation 2.26.

Regulation

The effect of factors such as device voltage drops, device forward resistance, conductor resistance and the AC source inductance, is to cause the converter output voltage on load (V_{load}) to differ from the ideal open circuit voltage V_{oc}. This difference is expressed by the regulation of the converter:

$$\text{Regulation} = \frac{V_{\text{oc}} - V_{\text{load}}}{V_{\text{oc}}} \times 100\% \qquad (2.28)$$

The open-circuit voltage (V_{oc}) in equation 2.28 is the ideal rectifier voltage with no overlap.

The voltage drop across a diode or thyristor may be represented by a constant value, or more accurately as in Fig. 2.25, by a combination of a constant voltage and a resistance. The precise values used must take account of the firing angle.

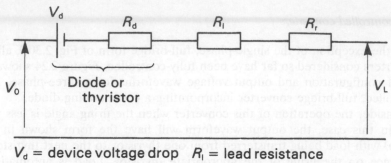

V_d = device voltage drop R_l = lead resistance
R_d = device resistance R_r = effective resistance due
                                          to overlap (voltage drop only)

Fig. 2.25 Rectifier equivalent circuit including device and lead components.

The resistance of the leads and the AC source can be taken as constant in most systems. Where the bridge operation causes current to flow in two phases simultaneously then the effective AC source resistance in the equivalent circuit will be the sum of the phase resistances.

Power factor

The general expression for power factor is

$$\text{Power factor} = \frac{\text{Mean power}}{V_{\text{RMS}} I_{\text{RMS}}} = \frac{\frac{1}{T} \int_0^T vi \, dt}{V_{\text{RMS}} I_{\text{RMS}}} \tag{2.29}$$

A converter has been shown to draw a non-sinusoidal current at supply frequency from the AC system and this current can therefore be represented by a fundamental component at supply frequency together with a series of harmonics. Assuming that the AC system voltage remains sinusoidal, then power will be delivered by the AC system at the fundamental frequency only. Therefore

$$\text{Power} = V_{1,\text{RMS}} \, I_{1,\text{RMS}} \cos\phi_1 \tag{2.30}$$

where $I_{1,\text{RMS}}$ is the RMS amplitude of the fundamental component of the AC system current;

        $V_{1,\text{RMS}}$ is the RMS amplitude of the fundamental component of the AC system voltage;

        ϕ_1 is the phase displacement between $V_{1,\text{RMS}}$ and $I_{1,\text{RMS}}$

Substituting this relationship in Equation 2.29 gives

$$\text{Power factor} = \frac{V_{1,\text{RMS}} \, I_{1,\text{RMS}} \cos\phi_1}{V_{1,\text{RMS}} \, I_{\text{RMS}}}$$

$$= \frac{I_{1,\text{RMS}}}{I_{\text{RMS}}} \cos\phi_1 = \mu\cos\phi_1 \tag{2.31}$$

$V_{1,\text{RMS}} = V_{\text{RMS}}$ for an undistorted sinewave and

$I_{\text{RMS}} = (I_{1,\text{RMS}} + I_{2,\text{RMS}}$

$+ I_{3,\text{RMS}} + I_{4,\text{RMS}}$

$+ ...)^{1/2}$

in which $\mu(= I_{1,\text{RMS}}/I_{\text{RMS}})$ is referred to as the current distortion factor and $\cos\phi_1$ as the displacement factor.

Whenever harmonic currents are present then the distortion factor μ will be less than 1, even if the fundamental components of current and voltage are in phase ($\cos\phi_1 = 1$). For a fully-controlled converter with a constant load current and ignoring overlap then ϕ_1 will be equal to the firing angle α.

This means that a converter must be supplied with reactive volt-amperes (VARs) by the AC system in order to operate. In the case of large converters, this usually means that a VAR source is provided at the converter rather than relying on the capacity of the AC system. In early installations this VAR source took the form of a **rotary compensator** but now more usually consists of a **static compensator** in which power semiconductor switches are used to switch capacitors in or out as required or to control a saturable reactor.

The situation for a half-controlled converter is more complicated and reference must be made to the relevant AC current waveforms.

A discussion of reactive volt-amperes (VARs) is given in Bradley, D.A. (1994). *Basic Electric Power and Machines*. Chapman and Hall.

Find the power factor for: (1) a fully controlled, single-phase bridge converter at firing angles of 30° and 60°; (2) a half-controlled, single-phase bridge converter at the same firing angles. In both cases overlap and device forward voltage drops are ignored and the load current is assumed to be constant.

(1) Referring to Fig. 2.3 for the fully controlled, single-phase bridge, since the load current is constant:

$$I_{RMS} = I_L$$

The mean load voltage is given by Equation 2.3:

$$V_{mean} = \frac{2}{\pi}\hat{V}\cos\alpha = \frac{2\sqrt{2}}{\pi}V_{RMS}\cos\alpha$$

The load power is then $V_{mean}I_L$ and

$$\text{Power factor} = \frac{2\sqrt{2}}{\pi}\cos\alpha$$

Since the load current is constant and overlap is ignored, $\cos\phi_1 = \cos\alpha$. Hence

$$\mu = I_{1,RMS}/I_{RMS} = \frac{2\sqrt{2}}{\pi} = 0.9003$$

and is independent of the firing angle

(i) $\alpha = 30°$
  Power factor = 0.7797

(ii) $\alpha = 60°$
  Power factor = 0.4502

(2) Referring to Fig. 2.3 for the half-controlled bridge, the thyristor current, and hence the supply current is discontinuous. The RMS current is found from

$$I_{RMS} = \left[\frac{1}{\pi}\int_{\alpha}^{\pi}I_L^2\,d\theta\right]^{1/2} = I_L\left[\frac{(\pi - \alpha)}{\pi}\right]^{1/2}$$

The mean load voltage is given by Equation 2.3:

$$V_{mean} = \frac{1}{\pi}\hat{V}(1 + \cos\alpha) = \frac{\sqrt{2}}{\pi}V_{RMS}(1 + \cos\alpha)$$

The mean load power is then $V_{mean}I_L$ and

$$\text{Power factor} = \frac{\sqrt{2}}{\pi}(1 + \cos\alpha)\left(\frac{\pi}{\pi - \alpha}\right)^{1/2}$$

Worked example 2.5

The amplitude of the fundamental component of current is obtained from its Fourier coefficients as

$$b_1 = \frac{I_L}{\pi}\left[\int_{-(\pi-\alpha)}^{0} -\sin(\omega t)\,\mathrm{d}(\omega t) + \int_{\alpha}^{\pi}\sin(\omega t)\,\mathrm{d}(\omega t)\right]$$

$$= 2I_L(1+\cos\alpha)/\pi$$

and

$$a_1 = \frac{I_L}{\pi}\left[\int_{-(\pi-\alpha)}^{0} -\cos(\omega t)\,\mathrm{d}(\omega t) + \int_{\alpha}^{\pi}\cos(\omega t)\,\mathrm{d}(\omega t)\right]$$

$$= 2I_L(\sin\alpha)/\pi$$

Supply current waveform for a two-phase, half-controlled bridge

Amplitude of fundamental $= (a_1^2 + b_1^2)^{1/2} = 2\sqrt{2}\,I_L(1+\cos\alpha)^{1/2}\pi$. RMS value of the fundamental is therefore $2I_L(1+\cos\alpha)^{1/2}/\pi$. Hence

$$\mu = \frac{2}{\pi}\left(\frac{\pi}{\pi-\alpha}\right)^{1/2}(1+\cos\alpha)^{1/2}$$

when from the expression for power factor above:

$$\cos\phi_1 = (1+\cos\alpha)^{1/2}/\sqrt{2}$$

(i) $\alpha = 30°$
Power factor $= 0.9201$
$\mu = 0.9526$
$\cos\phi_1 = 0.9659$

(ii) $\alpha = 60°$
Power factor $= 0.7397$
$\mu = 0.8541$
$\cos\phi_1 = 0.866$

Transformer rating

The need for special transformer winding configuration has been demonstrated already with various converter configurations. When selecting these transformers their rating under the particular operating conditions must be determined. This rating will in many cases be different for the primary and secondary windings. This may be contrasted with the conditions of normal transformer operation in which the rating will be the same for both windings.

The rating of any individual winding is obtained as the product of the RMS current through the winding and the RMS voltage across the winding.

Worked example 2.7

A three-phase, half-wave, uncontrolled rectifier is supplying a constant current of 25 A at 240 V to its load. The rectifier is supplied from the secondary of an interconnected star transformer, the primary of which is connected to a three-phase, 660 V (line) supply. Find the ratings of the transformer primary and secondary windings.

From Equation (2.4) with $\alpha = 0°$:

$$\hat{V} = \frac{250 \times \sqrt{2} \times \pi}{3 \times \sqrt{2}} = 290.2\,\text{V}$$

Hence

$$V_{RMS} = \hat{V}/\sqrt{2} = 205.2\,\text{V}$$

From Fig. 2.14 it can be seen that this voltage is obtained as the phasor sum of the voltages in each part of the secondary winding. These voltages are equal in magnitude ($\hat{V}_w$) and differing in phase by 60°.

Hence

$$V_{RMS} = \sqrt{3}\,\hat{V}_w/\sqrt{2} = \sqrt{3}\,V_{w,RMS}$$

where $V_{w,RMS}$ is the RMS value of the voltage in each part of the secondary winding.

Thus

$$V_{w.RMS} = V_{RMS}/\sqrt{3} = 118.4\,\text{V}$$

As the current in each of the secondary windings flows for a third of a cycle,

Equation 2.6

$$I_{2,RMS} = 25/\sqrt{3} = 14.43\,\text{A}$$

The primary winding voltage is obtained from the line voltage

$$V_{1,RMS} = 660/\sqrt{3} = 381\,\text{V}$$

The turns ratio between the primary winding and its associated secondary windings is then

$$n = 381/118.4 = 3.218$$

The primary current flows for two-thirds of a cycle and has an amplitude of

$$I_1 = 25/3.218 = 7.77\,\text{A}$$

The RMS value of the primary current is obtained from

$$I_{1,RMS} = \left[\frac{1}{2\pi}\left(\int_{-\frac{2\pi}{3}}^{0} I_L^2\,d\theta + \int_{\frac{\pi}{3}}^{0} I_L^2\,d\theta\right)\right]^{1/2} = \frac{\sqrt{2}}{3}\,I_L = \frac{\sqrt{2} \times 7.77}{3}$$

$$= 6.34\,\text{A}$$

The ratings are then

Primary $= 3 \times 6.34 \times 381 = 7.25\,\text{kW}$

Secondary $= 6 \times 14.43 \times 118.4 = 10.2\,\text{kW}$

Converters with discontinuous current

Where the load inductance is insufficient to maintain the DC current at a constant value, the output current will contain a ripple component which will be reflected in the supply current, as in Fig. 2.26(a) and (b). Under light load conditions this current may well become discontinuous as in Fig. 2.26(c). Analysis of both converter and load behaviour is now much more complex and requires that individual component values be taken into account.

The supply current will also be discontinuous when capacitor smoothing is used on the output of a rectifier as in Fig. 2.27(a). In this case, the diodes will

Arrillaga, J., Bradley, D.A. and Bodger, P.S. (1985). *Power System Harmonics.* John Wiley, U.K.

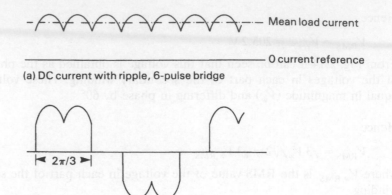
— Mean load current

— 0 current reference

(a) DC current with ripple, 6-pulse bridge

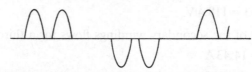

|← 2π/3 →|

(b) Phase current, 6-pulse bridge

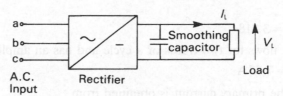

(c) Discontinuous phase current, 6-pulse bridge

Fig. 2.26 Current waveforms for a six-pulse bridge with a low inductance load.

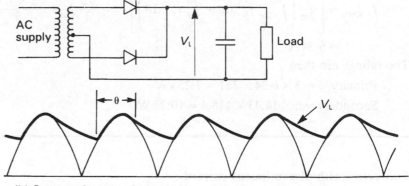

I_L

Smoothing capacitor

Load

V_L

A.C. Input

Rectifier

(a) Rectifier with capacitor smoothing.

AC supply

V_L

Load

|← θ →|

V_L

(b) Output of a two-phase, half-wave bridge with capacitor smoothing

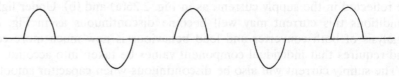

(c) Supply current waveform

Fig. 2.27 Operation of a rectifier with capacitive smoothing.

start to conduct once the AC voltage at their anode exceeds that of the smoothing capacitor. Conduction ceases when the AC anode voltage falls below the voltage on the smoothing capacitor. Figure 2.27(b) illustrates these conditions. As the load on the rectifier increases, the supply current will become more 'peaky' in order to supply the energy requirements of the load as suggested by Fig. 2.27(c) and (d). Conduction still only takes place over part of each half cycle.

With reference to Fig. 2.27(b) the conduction period θ can be approximated as

$$\theta = 2\cos^{-1}\left(V_{\mathrm{mean}}/\hat{V}\right) \tag{2.32}$$

Converters with voltage bias

When a converter is being used to supply a load such as a DC machine, the back EMF of the machine appears as a bias voltage on the DC side of the converter. The effect on the operation of a single-phase, fully-controlled converter is then as shown in Fig. 2.28. The performance of the converter under these conditions depends on the relationship between the firing angle α of the thyristors, the point-on-wave at which the AC source voltage exceeds the bias voltage ψ and the point-on-wave at which the thyristors stop conducting σ. Figures 2.28(b)–(e) illustrate one possible condition for the operation of the converter with $\alpha > \psi$ and a discontinuous load current when the effective load voltage is

$$v_{\mathrm{L}} = \hat{V}\sin\omega t \qquad\quad \text{for } \alpha < \omega t < \sigma$$

and

$$v_{\mathrm{L}} = E_{\mathrm{a}} \qquad\qquad \text{for } (\sigma - \pi) < \omega t < \alpha$$

Therefore

$$
\begin{aligned}
V'_{\mathrm{L}} &= \frac{1}{\pi}\left[\int_{\sigma-\pi}^{\alpha} E_{\mathrm{a}}\, \mathrm{d}(\omega t) + \int_{\alpha}^{\sigma} \hat{V}\sin\omega t\, \mathrm{d}(\omega t)\right] \\
&= \frac{1}{\pi}\left[\hat{V}(\cos\alpha - \cos\sigma) + E_{\mathrm{a}}(\alpha + \pi - \sigma)\right]
\end{aligned}
\tag{2.33}
$$

A discussion of DC machines is given in Chapter 4.

Dewan, S.B., Slemon, G.R. and Straughen, A. (1984). *Power Semiconductor Drives.* John Wiley.

α, ψ and σ are all measured from the same voltage zero.

The thyristors cannot be fired at $\alpha < \psi$.

Problems

2.1 A circuit such as that of Fig. 2.2(b) is supplied from a 50 Hz supply via a transformer such that

$$V_{1,\mathrm{RMS}} = V_{2,\mathrm{RMS}} = 220\,\mathrm{V}$$

Neglecting any voltage drop in the thyristors, find the mean load current at firing angles of 30° and 60° if the load is a pure resistance of 15 Ω. What will be the RMS and peak current in the thyristors under each of the above conditions?

If an inductance of 18 mH is included in series with the resistive load, estimate the firing angle at which the load current will become continuous.

2.2 A three-phase, half-wave converter is supplying a load with a continuous, constant current of 40 A over a range of firing angles from 0° to

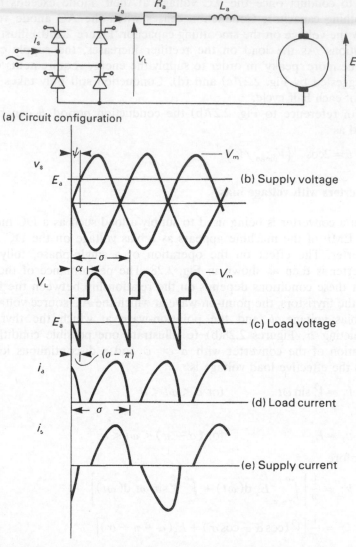

(a) Circuit configuration

(b) Supply voltage

(c) Load voltage

(d) Load current

(e) Supply current

Fig. 2.28 Single-phase fully-controlled converter with DC machine load. Firing angle (α) greater than cut-off angle (ψ) and ψ greater than ($\sigma - \pi$).

75°. What will be the power dissipated by the load at these limiting values of firing angle? The supply voltage is 415 V (line).

2.3 A single-phase, half-controlled bridge is constructed as in the accompanying figure. Sketch the load voltage waveform for an inductive load at a firing angle of 60°.

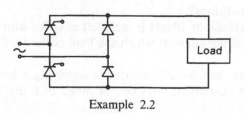

Example 2.2

2.4 A half-controlled, single-phase bridge with a commutating diode as shown in Fig. 2.3(b) is fed by a 100 V, 50 Hz source and is supplying a constant load current of 20 A at a firing angle of 60°. If the source has an inductance of 0.34 H, find the overlap angles when the thyristor (a) turns on and (b) turns off.

2.5 A three-phase, fully-controlled bridge converter is fed from an inter-connected star transformer as in Fig. 2.6 and is supplying a highly inductive load of resistance 8 Ω. The transformer provides a secondary phase voltage of 240 V from a primary phase voltage of 660 V. Determine the rating of the transformer. Ignore overlap and thyristor voltage drop.

2.6 A three-phase, fully-controlled bridge converter is supplying a DC load of 400 V, 60 A from a three-phase, 50 Hz, 660 V (line) supply. If the thyristors have a forward voltage drop of 1.2 V when conducting then, ignoring overlap, find (a) the firing angle of the thyristors, (b) the RMS current in the thyristors and (c) the mean power loss in the thyristors.

If the AC supply has an inductance per phase of 3.6 mH, what will be the new value of firing angle required to meet the load requirements?

2.7 A DC load with a maximum rating of 100 kV, 500 A is to be supplied by a 12 pulse bridge converter made up of two bridge converters as in Fig. 2.23. Neglecting overlap and thyristor voltage drops, determine the required thyristor ratings for (a) parallel connection and (b) series connection of the bridges.

2.8 A three-phase, fully-controlled bridge converter is connected to a three-phase, 50 Hz, 415 V (line) supply and is operating in the inverting mode at a firing advance angle of 30°. If the AC supply has a resistance and inductance per phase of 0.04 Ω and 1 mH respectively, find the DC source voltage, overlap angle and recovery angle when the DC current is constant at 52 A.

The thyristors have a forward voltage drop when conducting of 1.8 V.

2.9 For the system of Problem 2.8, what will be the maximum DC current that can be accommodated at a firing advance angle of 22.5°, allowing for a recovery angle of 5°?

3
DC choppers, inverters and cycloconverters

□ To examine the operation of thyristors with a DC source voltage.
□ To introduce forced commutation and to consider the operation of various forced-commutation circuits.
□ To consider the operation of a DC chopper.
□ To classify DC choppers in terms of their operating envelopes.
□ To examine the operation of voltage- and current-sourced converters as a means of producing a variable-frequency supply.
□ To consider means of controlling a variable frequency inverter output voltage.
□ To introduce pulse-width-modulated inverters.
□ To introduce the cycloconverter.

Objectives

The symbol adopted for a thyristor with forced commutation is:

The following symbol is used in the text for a thyristor with external commutating circuits:

Silicon Controlled Rectifier Manual, General Electric, New York.

Chapter 2 described a range of circuits capable of producing a DC voltage from an AC source voltage. In each of these circuits, the action of the AC source voltage resulted in the conducting diode or thyristor being turned off at a natural current zero or on the transfer of load current to another diode or thyristor.

However, there also exists a large class of power electronic circuits such as chopper regulators and variable frequency inverters which operate from a DC source and rely on control of both the turn-on and turn-off of the switching device for their operation. Where self-commutating devices such as power transistors, power MOSFETs, IGBTs and GTO thyristors are used, turn-off can be achieved by control of the device base or gate conditions. In applications where thyristors are required for reasons of rating then the achievement of controlled turn-off requires the use of external circuits and is referred to as **forced commutation.**

Forced commutation circuits

The function of a forced commutation circuit is first to reduce the current through the conducting thyristor to zero and then to maintain a reverse voltage across this thyristor for a period equal to or greater than the thyristor turn-off time in order to re-establish the blocking state.

Commutation by an external voltage source

The above objectives can be met by using external voltage source to reduce the forward current through the thyristor to zero by driving a reverse current

71

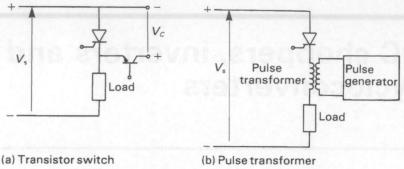

(a) Transistor switch   (b) Pulse transformer

Fig. 3.1 Commutation by an external voltage source.

of sufficient magnitude through the thyristor. The external voltage source then provides and maintains the necessary reverse voltage conditions across the thyristor to complete the turn-off.

Figure 3.1 shows two means by which such an external voltage can be applied to a thyristor. In Fig. 3.1(a) a transistor switch is used to place an appropriate external voltage source across the thyristor while in Fig 3.1(b), a pulse transformer is used to introduce the commutating voltage into the main circuit.

Commutation using a capacitor

The reverse voltage could also be applied using the arrangement of Fig. 3.2. With the capacitor initially charged in the direction shown in the figure, when the switch (S) is closed, the capacitor will begin to discharge through the thyristor, forcing the forward current below the level of the holding current. The capacitor will then continue to discharge through the load, maintaining the necessary reverse voltage across the thyristor to complete the turn-off before recharging in the reverse direction.

In practice, additional circuitry must be provided to control the charging of the capacitor and to initiate the discharge. Figure 3.3(a) shows a simple forced commutation circuit using a second or commutating thyristor to provide the control while Fig. 3.3(b) shows the voltage and current waveforms at various points in the circuit during turn-off and turn-on.

The reduction of the forward current to zero takes place almost instantaneously.

With the main thyristor T_1 conducting at time t_0, the capacitor is charged in the direction shown in the figure. When the commutating thyristor T_2 is fired at time t_1, the capacitor is connected across the main thyristor, forcing a

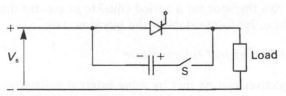

Fig. 3.2 Commutation by a parallel capacitor.

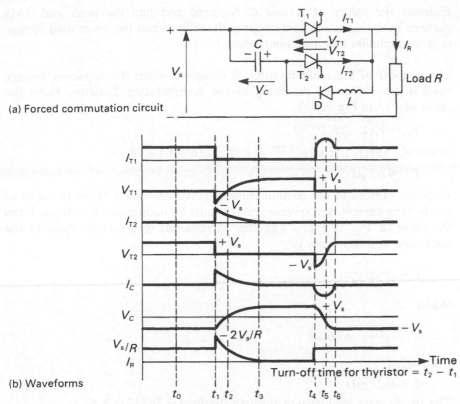

(a) Forced commutation circuit

(b) Waveforms

Turn-off time for thyristor = $t_2 - t_1$

Fig. 3.3 Commutation by an auxiliary capacitor.

reverse current through this thyristor and reducing the forward current to zero. Thyristor T_2 will then continue to conduct as the capacitor continues to discharge through the load, maintaining a reverse current across the main thyristor until time t_2 at which point the turn-off of this thyristor is complete. The capacitor will then continue to charge in the reverse direction until time t_3 at which point the forward current through the commutating thyristor T_2 falls below the holding current.

When the main thyristor is next fired at time t_4, the capacitor will begin to discharge via thyristor T_1, diode D and the inductance L. During the first period of this oscillatory discharge, until time t_5, a reverse voltage is placed across the commutating thyristor T_2, thus completing its turn-off. After one half-cycle of the oscillation, at time t_6, the current through the diode attempts to reverse and the diode ceases to conduct. The capacitor is now charged in the original direction of the figure ready for the next commutation cycle.

The frequency of the oscillation is determined by the values of the capacitor and inductor since $\omega = 1/\sqrt{(LC)}$.

In the circuits of Figs 3.2 and 3.3 the thyristors are connected to the high-voltage side of the load and the gate must therefore be isolated using techniques such as those discussed in Chapter 1.

The circuit of Fig. 3.3(a) is being used to supply a 12 Ω load from a 36 V DC source. The switching frequency is 250 Hz and the main and commutating thyristors each have a holding current of 50 mA and a turn-off time of 80 μs.

Worked example 3.1

73

Estimate the values of L and C required and find the peak and RMS currents in the main and commutating thyristors when the mean load voltage is at its minimum and maximum values.

The turn-off of the main thyristor is complete when the capacitor voltage reaches zero following the firing of the commutating thyristor. From the curve of V_{T1} in Fig. 3.3(b):

$$v_{T1} = V_s(1 - 2e^{-t/RC})$$

When V_{T1} reaches zero, $e^{-t/RC} = 1/2$, when for $t = 80$ μs:

$$C = 9.62 \ \mu F$$

Once the current in the commutating thyristor has fallen below the level of the holding current, the reverse voltage must be maintained for 80 μs. From the curve of V_{T2} in Fig. 3.3(b), this corresponds to a quarter cycle of the oscillation at a frequency of

$$f = \frac{1}{2\pi \sqrt{(LC)}}$$

When

$$\frac{2\pi \sqrt{(LC)}}{4} = 80 \ \mu s$$

from which

$$L = 0.27 \text{ mH}$$

The steady state load current through the load $= 36/12 = 3$ A.

The peak capacitor current may be obtained from energy considerations using

$$\frac{1}{2}CV^2 = \frac{1}{2}LI^2$$

when, for a lossless system in which the capacitor voltage is the supply voltage V_s:

$$I_{C,max} = V \sqrt{\left(\frac{C}{L}\right)} = 6.8 \text{ A}$$

which must be supplied by the main thyristor.

The peak current in the main thyristor is then

$$6.8 + 3 = 9.8 \text{ A}$$

The peak current in the commutating thyristor is the capacitor current immediately after firing when

$$I_{T2,max} = 2V_s/R = 6 \text{ A}$$

The commutating thyristor stops conducting when the forward current falls below 50 mA, therefore

$$50 \times 10^{-3} = 6e^{-t/RC}$$

when

$$t = 553 \ \mu s$$

which is the minimum turn-off time for the commutating thyristor. If the main thyristor is fired before this then the commutating thyristor will be forced off but the capacitor will not be fully charged. Usually, the off period is made long enough to ensure that the voltage on the capacitor has reached at least 80% of its maximum value.

The minimum on time for the main thyristor is therefore normally equivalent to a half cycle of the oscillatory waveform of 160 μs.

Repetition period $= 1/250 = 4$ ms

Minimum load voltage corresponds to the minimum on time of the main thyristor when

$$I_{T1,RMS} = \left(\frac{1}{4 \times 10^{-3}} \int_0^{160 \times 10^{-6}} (3 + 6.8 \sin \omega t)2 \, dt \right)^{1/2} = 1.13 \text{ A}$$

and

$$I_{T2,RMS} = \left(\frac{1}{4 \times 10^{-3}} \int_0^{160 \times 10^{-6}} (6e^{-t/RC})2 \, dt \right)^{1/2} = 0.52 \text{ A}$$

The maximum mean load voltage corresponds to the maximum on period of the main thyristor when

$$I_{T1,RMS} = \left(\frac{1}{4 \times 10^{-3}} \int_0^{160 \times 10^{-6}} (3 + 6.8 \sin \omega t)2 \, dt \right.$$
$$\left. + \int_{160 \times 10^{-6}}^{3.467 \times 10^{-3}} 32 \, dt \right)^{1/2}$$
$$= 2.95 \text{ A}$$

$I_{T2,RMS}$ remains the same as before.

The Jones circuit of Fig. 3.4 also uses an auxiliary thyristor for commutation purposes and has the advantage over the circuit of Fig. 3.3(a) of reliably initiating the commutation sequence following the initial turn-on with the capacitor uncharged. This occurs since with the capacitor uncharged then on firing the main thyristor T_1 the coupling between the inductors L_1 and L_2

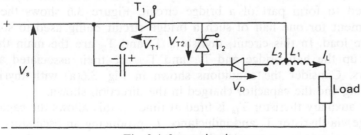

Fig. 3.4 Jones circuit.

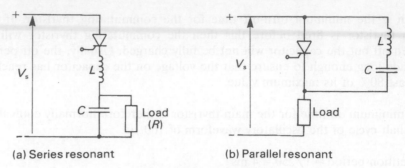

(a) Series resonant   (b) Parallel resonant

Fig. 3.5 Resonant turn-off.

causes a voltage to be induced into inductance L_2 which charges the capacitor via diode D in the direction shown in the figure. The capacitor is now charged in the correct direction to turn off the main thyristor when the commutating thyristor T_2 is fired.

Subsequent operation of the circuit of Fig. 3.4 follows the pattern of Fig. 3.3.

Resonant turn-off

The resonance set up in an auxiliary LC circuit can also be used to turn a thyristor off, eliminating the need for auxiliary thyristors or diodes, though at the expense of a loss of control over the instant of turn-off. Fig. 3.5(a) shows a simple series resonant turn-off ciruit. Provided this circuit, including the load, is underdamped, then firing the thyristor will set up a current oscillation which will turn the thyristor off at the first current zero. The capacitor will then discharge through the load. The on time of the thyristor is determined by the frequency of the oscillatory circuit.

Parallel resonance circuits such as that shown in Fig. 3.5(b) can also be used for thyristor turn-off. In this circuit, the capacitor is charged during the thyristor off period to the supply voltage. When the thyristor is fired, an oscillatory current is set up which, provided it is greater than the supply current, will turn the thyristor off. The thyristor on period is again a function of the frequency of oscillation of the LC circuit, while the off period must be sufficient to allow the capacitor to be adequately charged.

Bridge circuits

More complex commutation circuits may be required when the thyristors are connected to form part of a bridge circuit. Figure 3.6 shows the circuit arrangement for one half of such a bridge circuit being used to supply an inductive load. In this circuit, thyristors T_1 and T_2 are the main thyristors making up the half bridge and T_{1a} and T_{2a} are their associated auxiliary thyristors. Consider the conditions shown in Fig. 3.6(a) with thyristor T_1 conducting and the capacitor charged in the direction shown.

If the auxiliary thyristor T_{1a} is fired at time t_0, this allows the capacitor to discharge via thyristor T_1 and inductance L, producing an oscillatory current whose magnitude is arranged to be much greater than the load current. As

The capacitor is usually charged to at least $0.8V_s$.

The McMurray circuit.

Murphy, J.M.D. (1973). *Thyristor Control of AC Motors.* Pergamon, Oxford.

Assumes an inductive load with I_L essentially constant over the commutation interval.

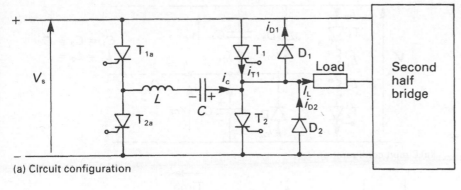

(a) Circuit configuration

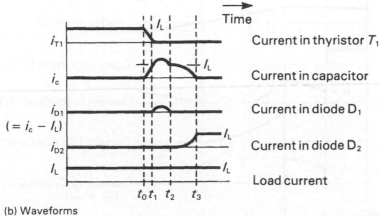

(b) Waveforms

Fig. 3.6 McMurray circuit.

this current initially flows through thyristor T_1 in the reverse direction, the forward current through this thyristor is reduced to zero at time t_1, at which point the oscillatory current is diverted to diode D_1. The forward voltage across D_1 now appears as a reverse voltage across thyristor T_1 to complete the turn-off and re-establish the blocking mode.

As the capacitor discharges, its voltage reverses with the discharge current falling below the load current at time t_2, at which point diode D_1 stops conducting. Load current continues to flow via the auxiliary thyristor T_{1a}, charging the capacitor to the source voltage in the reverse direction at which point diode D_2 starts to conduct. This diode then continues to take an increasing proportion of the load current as the energy in the magnetic field of the inductor L is transferred to the capacitor. This reduces the current in the auxiliary thyristor T_{1a} to zero to turn it off. The capacitor is now charged to approximately $2V_s$ in the reverse direction ready for the commutation of thyristor T_2 by the auxiliary thyristor T_{2a}.

The load current now decays to zero, turning diode D_2 off and allowing thyristor T_2 to start conducting, reversing the current in the load. T_2 would normally be supplied with a continuous gate signal during this period to ensure turn-on at the earliest appropriate instant.

A variation on the circuit of Fig. 3.6 is shown in Fig. 3.7. Initially, with thyristor T_1 conducting, capacitor C_1 is uncharged and capacitor C_2 charged

The opposite half-bridge is assumed to have been commutated simultaneously.

McMurray–Bedford circuit.

$L_1 = L_2 = L$ and $C_1 = C_2 = C$

77

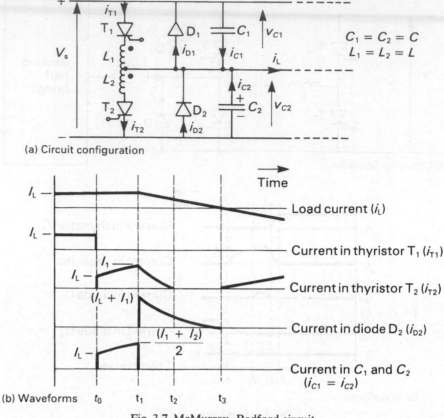

(a) Circuit configuration

$C_1 = C_2 = C$
$L_1 = L_2 = L$

(b) Waveforms

Fig. 3.7 McMurray–Bedford circuit.

as shown. When thyristor T_2 is fired at time t_0, one end of inductor L_2 is connected to the negative rail of the supply. Since the capacitors cannot change voltage instantaneously, the supply voltage now appears across L_2. As inductors, L_1 and L_2 are close coupled, an equivalent voltage is induced in L_1, raising the cathode potential of thyristor T_1 to $2V_s$ to turn it off.

The load current now transfers to thyristor T_2 and inductor L_2, preserving the ampere-turns balance in the L_1L_2 coil and maintaining the reverse bias on thyristor T_1. The current in the inductive load is maintained during this period by the charging currents of capacitors C_1 and C_2. As capacitor C_2 discharges, the voltage across inductor L_2 is decreased, reducing the induced voltage in L_1. At the same time, capacitor C_1 is charging, eliminating the reverse bias on thyristor T_1 when the forward voltage on capacitor C_1 exceeds the reverse voltage on inductor L_1.

The load current transfers to diode D_2 at time t_1, at which point capacitor C_1 is charged to V_s. The energy stored in inductor L_2 is then dissipated in the loop L_2–T_2–D_2, with the current in thyristor T_2 falling to zero at time t_2. Diode D_2 continues to supply a decreasing current, turning off when the load current reaches zero at time t_3. A reverse load current can now be supplied via thyristor T_2.

Thyristor T_2 turns off when the energy stored in L_2 has been dissipated.

Thyristor T_2 would be supplied with a continuous gate signal to ensure turn-on at the appropriate instant.

A full bridge circuit of the form shown in Fig. 3.6 is used to supply an inductive load such that the load current is continuous during commutation. The load resistance is 12 Ω and the bridge is supplied from a constant 400 V DC source. The thyristors used in the bridge have a turn off time of 50 μs. Estimate the necessary component values.

Referring to Fig. 3.6(b), it is seen that a reverse voltage equal to the forward voltage drop of the parallel diode appears across the main thyristor in the period t_1 to t_2. This interval must therefore correspond to the turn-off time of the thyristor.

From Fig. 3.6(b), this period can be seen to be equal to the period for which the capacitor current I_c is greater than the load current. As I_c has the form

$$i_C = \hat{I}_C \sin \omega_0 t$$

where $\omega_0 = 1/\sqrt{(LC)}$

Then, by reference to Fig. 3.6(b):

$$I_L = \hat{I}_C \sin \omega_0 t_1 \quad (\omega_0 t_1 < \pi/2)$$

and

$$I_L = \hat{I}_C \sin \omega_0 t_2 \quad (\omega_0 t_2 > \pi/2)$$

Turn-off time $= 50 \ \mu s = t_2 - t_1$

Energy available for turn-off $= \frac{1}{2} C V_C^2 = \frac{1}{2} L \hat{I}_C^2$

Let $V_c = 2 \times 400 = 800$ V

Setting $\hat{I}_C = 1.5 I_L = \dfrac{1.5 \times 400}{12} = 50$ A

Then, from above:

$$\omega_0 t_1 = 0.7297 \text{ and } \omega_0 t_2 = 2.4119$$

Thus

$$t_2 - t_1 = 1.6822/\omega_0 = 50 \ \mu s$$

when

$$\omega_0 = \frac{1.6822}{50 \times 10^{-6}} = 3.364 \times 10^4 \text{ radians s}^{-1} = \frac{1}{\sqrt{(LC)}}$$

From the energy relationship

$$\frac{C}{L} = \frac{\hat{I}_C^2}{V_C^2} = \frac{50^2}{800^2} = 3.906 \times 10^{-3}$$

when

$$C = 3.906 \times 10^{-3} L$$

Substituting in the expression for ω_0 gives

$$L = 0.476 \text{ mH and } C = 1.86 \text{ } \mu\text{F}$$

DC choppers

When the GTO thyristor in Fig. 3.8 is turned off, current continues to flow in the load via the freewheeling diode.

A DC chopper is used to provide a controllable DC output from a DC source by switching the source on to and off the load. By varying the switching frequency with a constant ON period or the *mark–space ratio* at constant frequency, the voltage at the load can be controlled. Figure 3.8(a) shows a basic step-down chopper regulator using a GTO thyristor and supplying an inductive load. The mean load voltage is

$$V_L = V_s \, t_1 / T \tag{3.1}$$

while the RMS voltage at the load is given by

$$V_{L.RMS} = V_s \sqrt{(t_1/T)} \tag{3.2}$$

Load time constant $\tau = L/R$.

For linear variation $L/R > 10\,T$

Bird, B.M., King, K.G. and Pedder, D.A.G. (1993). *An Introduction to Power Electronics*. John Wiley. U.K.

With a load inductance such that the time constant of the load is much greater than the switching interval (T), the waveforms of load current and load voltage will be of the form shown in Figs 3.8(b) and 3.8(c) respectively. Under these conditions, the variation of the load current may be assumed to be linear, in which case, during the conduction period of the GTO thyristor

$$V_s - V_L = L \frac{di}{dt} = L \frac{\Delta i}{\Delta t} \tag{3.3}$$

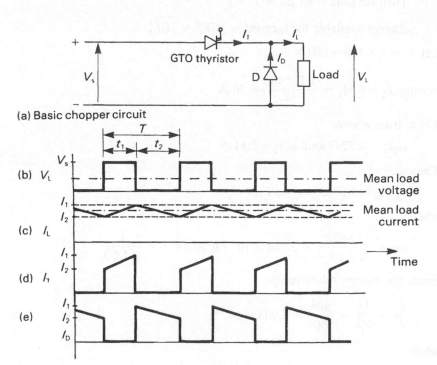

(a) Basic chopper circuit

Fig. 3.8 Operation of a basic chopper with smoothing or T much less than system time constant.

80

and

$$I_1 - I_2 = \frac{(V_s - V_L)t_1}{L} \tag{3.4}$$

During the OFF period

$$I_1 - I_2 = \frac{V_L(T - t_1)}{L} = \frac{t_2 V_L}{L} \tag{3.5}$$

Also

$$I_{mean} = (I_1 + I_2)/2 = V_L/R \tag{3.6}$$

Hence, from Equations 3.5 and 3.6

$$I_1 = I_{mean} + \frac{t_2 V_L}{2L} \tag{3.7}$$

and

$$I_2 = I_{mean} - \frac{t_2 V_L}{2L} \tag{3.8}$$

The ripple current (i_r) can now be expressed as

$$i_r = (I_1 - I_2)\left(\frac{t}{t_1} - \frac{1}{2}\right) \text{ for } 0 < t < t_1 \tag{3.9}$$

and

$$i_r = (I_1 - I_2)\left(\frac{1}{2} - \frac{(t - t_1)}{t_2}\right) \text{ for } t_1 < t < T \tag{3.10}$$

The RMS value of the ripple current is then

$$I_{r,RMS} = \left\{\frac{(I_1 - I_2)^2}{T}\left[\int_0^{t_1}\left(\frac{t}{t_1} - \frac{1}{2}\right)^2 dt + \int_{t_1}^{T}\left(\frac{1}{2} - \frac{(t - t_1)}{t_2}\right)^2 dt\right]\right\}^{1/2}$$

$$= \frac{(I_1 - I_2)}{2\sqrt{3}} \tag{3.11}$$

The mean load current can be obtained by dividing the mean load voltage (V_L) by the load resistance.

If the period T is of the order of the load time constant then, in the absence of any other form of smoothing, the variation of the load current can no longer be considered to be linear. With reference to Fig. 3.9, during the $L/R < 10\,T.$

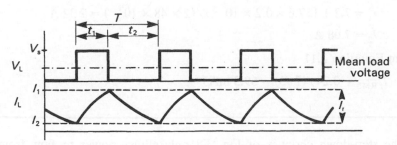

Fig. 3.9 Operation of a basic chopper with unsmoothed output where T is of the order of the system time constant.

conduction period of the GTO thyristor

$$i_L = I_2 + \left(\frac{V_s}{R} - I_2\right)(1 - e^{-Rt/L}) \tag{3.12}$$

and, during the OFF period

$$i_L = I_1 e^{-Rt/L} \tag{3.13}$$

Worked example 3.3

A DC chopper of the type shown in Fig. 3.8 is operating at a frequency of 2 kHz from a 96 V DC source to supply a load of resistance 8 Ω. The load time constant is 6 ms. If the mean load voltage is 57.6 V, find the mark–space ratio of the voltage waveform, the mean load current, the magnitude of the ripple current and its RMS value.

Period $= T = 1/f = 1/2000 = 0.5$ ms

Load time constant $= 12T$, therefore assume a linear variation of load current.

From Equation 3.1,

$$V_L = 57.6 = 96t_1/T$$

$$\therefore \quad t_1 = 0.3 \text{ ms}$$

From Equation 3.2,

$$V_{L,RMS} = 96 \times (0.3/0.5)^{1/2} = 74.36 \text{ V}$$

Mean load current is then found from the mean load voltage and the load resistance:

$$I_{mean} = 57.6/8 = 7.2 \text{ A}$$

Load time constant $= L/R$, therefore

$$L = 6 \times 10^{-3} \times 8 = 48 \text{ mH}$$

From Equation 3.3,

$$\text{Current ripple} = \Delta i = (V_s - V_L)\Delta t/L$$

$$= \{(96 - 57.6) \times 0.3 \times 10^{-3}\}/(48 \times 10^{-3}) = 0.24 \text{ A}$$

From Equation 3.7

$$I_1 = 7.2 + (57.6 \times 0.2 \times 10^{-3})/(2 \times 48 \times 10^{-3}) = 7.32 \text{ A}$$

$$\therefore I_2 = 7.08 \text{ A}$$

From Equation 3.11

$$I_{r,RMS} = 0.24/(2\sqrt{3}) = 69.3 \text{ mA}$$

Dewan, S.B., Slemon, G.R. and Straughen, A. (1984). *Power Semiconductor Drives*, John Wiley, New York.

The step-down chopper of Fig. 3.8 only allows power to flow from the supply to the load and is referred to as a class A or single quadrant chopper

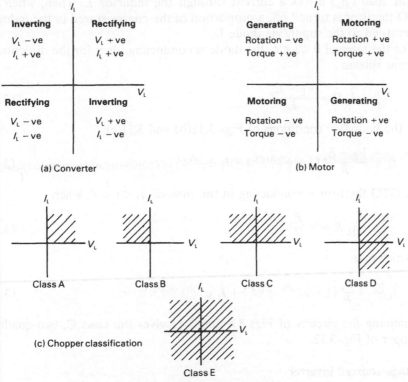

(a) Converter

(b) Motor

(c) Chopper classification

Fig. 3.10 Quadrant diagrams.

as it operates only in the first quadrant of the V_L/I_L diagram of Fig. 3.10(a). Other chopper circuits capable of operating in one, two or four quadrants are classified according to Fig. 3.10(c).

Figure 3.11(a) shows a class B, step-up chopper using a GTO thyristor as the switching element. When the GTO thyristor is turned on, the back EMF

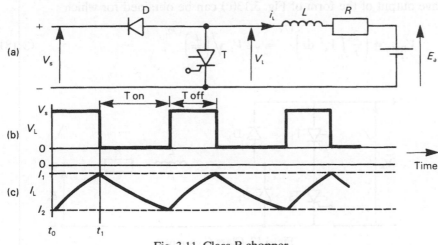

Fig. 3.11 Class B chopper.

of the load (E_a) drives a current through the inductor L. Then, when the GTO thyristor is turned off, a proportion of the energy stored in the inductor is returned to the supply via diode D.

For the interval $0 < t < t_1$ the diode is conducting, then for the direction of current shown:

$$\frac{di_L}{dt} + i_L R = \frac{V_L - E_a}{L} \tag{3.14}$$

Figures 3.11(b) and 3.11(c) represent steady-state operation.

For the boundary conditions of Figs 3.11(b) and 3.11(c):

$$i_L = \frac{V_L - E_a}{R}(1 - e^{-Rt/L}) + I_2\, e^{-Rt/L} \tag{3.15}$$

The GTO thyristor is conducting in the interval $t_1 < t < T$, when

$$\frac{di_L}{dt} + i_L R = -\frac{E_a}{L} \tag{3.16}$$

in which case

$$i_L = -\frac{E_a}{R}(1 - e^{-R(t - t_1)/L}) + I_1\, e^{-R(t - t_1)/L} \tag{3.17}$$

Care must be taken with the circuit of Fig. 3.12 to ensure that the GTO thyristors T_1 and T_2 are not fired together as this would short-circuit the supply.

Combining the circuits of Figs 3.8 and 3.11 gives the class C, two-quadrant chopper of Fig. 3.12.

Voltage-sourced inverter

Figure 3.13 is a single-phase, voltage-sourced inverter. The reverse diodes across the GTO thyristors accommodate the phase difference between the current and voltage in an inductive load.

Two class C, two-quadrant choppers can be combined as in Fig. 3.13(a) to give a class E full four-quadrant chopper circuit. This circuit is capable of producing a bidirectional (alternating) current in the load and is the basis of the voltage-sourced inverter.

Induction motor operation and control is discussed in Chapter 4.

If the pairs of GTO thyristors T_1–T_2 and T_3–T_4 are fired at equal intervals, the load voltage will be the square wave of Fig. 3.13(b). However, for applications such as induction motor drives, some control of the amplitude of the inverter output voltage as well as output frequency is required. By delaying the firing of one pair of thyristors relative to the other, a quasi-square wave output of the form of Fig. 3.13(c) can be obtained for which

$$V_{RMS} = \left(\frac{2}{T} \int_0^t V_s^2\, dt \right)^{1/2} = \sqrt{2}\, V_s \sqrt{\left(\frac{t}{T} \right)^{1/2}} \tag{3.18}$$

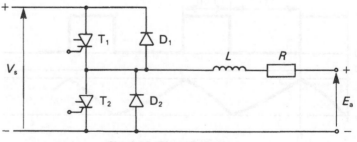

Fig. 3.12 Class C chopper.

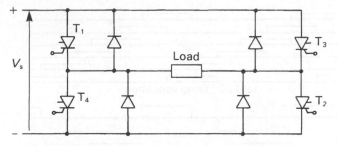

(a) Basic circuit, excluding commutation circuits

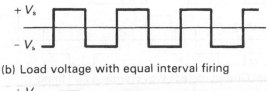

(b) Load voltage with equal interval firing

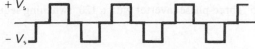

(c) Quasi-square wave output produced by delayed firing

Fig. 3.13 Voltage-sourced single-phase bridge inverter.

Three-phase bridge inverter

A three-phase, voltage-sourced bridge inverter can be produced by combining three single-phase half bridges as in Fig. 3.14 in which transistors are used as the switching device. An inverter of this type is typically operated with either a 120° or a 180° conduction sequence for the switching devices.

With a 120° switching sequence and assuming a star-connected load, the load voltage waveforms will be as shown in Fig. 3.15. If instead, the 180° switching sequence is used the load voltage waveforms will be as shown in Fig. 3.16, including a triple frequency neutral voltage.

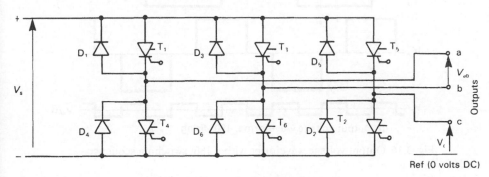

Fig. 3.14 Three-phase bridge inverter.

85

The effective instantaneous connection for a star-connected load with a 120° switching sequence is

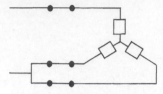

The star point of the load is at the zero reference with a 120° switching sequence. Voltages V_a, V_b and V_c in Fig. 3.15 are measured with respect to the star point.

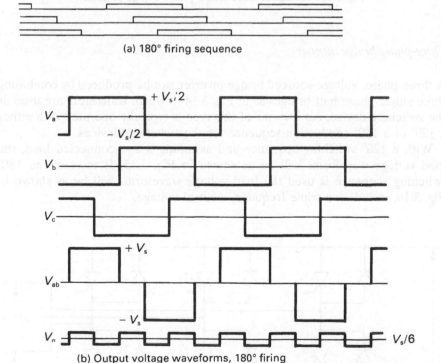

Fig. 3.15 Three-phase inverter with a 120° switching sequence.

The effective instantaneous connection for a star-connected load with a 180° switching sequences is

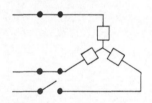

Voltages V_a, V_b, V_c and V_n in Fig. 3.16 are all measured with respect to the zero reference. Relative to the neutral, the phase 'a' voltage has the form

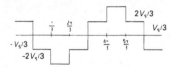

Fig. 3.16 Output voltage waveforms with a 180° switching sequence.

The use of self-commutating devices such as GTO thyristors, transistors, power MOSFETs and IGBTs as the switching element in the voltage-source bridge inverter removes the need for any forced commutation circuitry. Inverters using self-commutating devices also tend to be capable of operating at much higher switching rates than inverters using conventional thyristors and are therefore more suited to use in applications such as pulse-width modulated (PWM) inverters.

A three-phase bridge inverter such as that shown in Fig. 3.14 is supplied from a 600 V DC source. For a star-connected resistive load of 15 Ω per phase find the RMS load current, the load power and the device current ratings for (a) 120° conduction and (b) 180° conduction.

Worked example 3.4

(a) 120° conduction

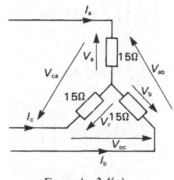

Example 3.4(a)

Load current amplitude $= 600/(2 \times 15) = 20$ A

$$\text{RMS load current} = \left\{ \frac{1}{2\pi} \left[\int_0^{\frac{2\pi}{3}} 20^2 \, d\theta + \int_\pi^{\frac{5\pi}{6}} 20^2 \, d\theta \right] \right\}^{1/2}$$

$$= [(20^2 + 20^2)/3]^{1/2} = 16.33 \text{ A}$$

Load power $= 16.33^2 \times 15 \times 3 = 12$ kW
Thyristor RMS current $= (20^2/3)^{1/2} = 11.5$ A

(b) 180° conduction
At any instant, load on inverter $= 15 + 15/2 = 22.5$ Ω

$I_1 = V_s/22.5 = 600/22.5 = 26.67$ A
$I_2 = I_1/2 = 13.33$ A

Phases are connected in parallel for two-thirds of a cycle, therefore

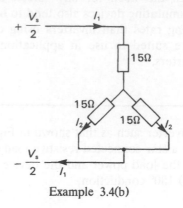

Example 3.4(b)

RMS load current =

$$\left\{\frac{1}{2\pi}\left[\int_0^{\frac{2\pi}{3}} 13.33^2 \, d\theta\right.\right.$$

$$+ \int_{\frac{2\pi}{3}}^{\frac{4\pi}{3}} 26.67^2 \, d\theta$$

$$\left.\left.+ \int_{\frac{4\pi}{3}}^{2\pi} 13.33^2 \, d\theta\right]\right\}^{1/2}$$

$$= \left(\frac{2 \times 13.33^2 + 26.67^2}{3}\right)^{1/2} = 18.85 \text{ A}$$

Thyristors carry a current of 26.67 A for one-sixth of a cycle and 13.33 for a half cycle. Therefore the RMS current in a thyristor is

RMS load current$/2 = 13.33$ A
Load power $= 18.85^2 \times 15 \times 3 = 15.99$ kW

Transformer-coupled inverter

Figure 3.17 shows a single-phase, transformer-coupled inverter. By alternately firing and turning off the thyristors T_1 and T_2, the supply voltage is connected to each half of the transformer primary winding in turn, producing an alternating voltage in the secondary winding. The two halves of the primary winding must themselves be close coupled in order to sustain the transfer load current from one half of the primary to the other at the turn-off of an individual GTO thyristor. By connecting the secondaries of two such inverters in series as in Fig. 3.18, a quasi-squarewave output can be obtained by varying the relative firing instants of the two bridges.

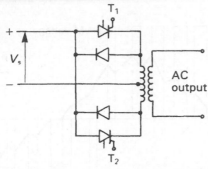

Fig. 3.17 Single-phase transformer-coupled inverter.

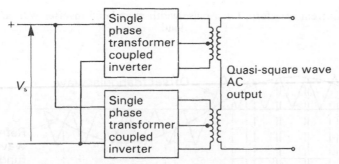

Fig. 3.18 Transformer-coupled inverters connected in series to produce a quasi-squarewave output.

Pulse-width-modulated inverters

In the simplest form of pulse-width-modulated (PWM) inverter, the supply voltage is switched at regular intervals to produce a load voltage waveform of the form shown in Fig. 3.19. Control of the load voltage amplitude is achieved by varying the mark–space ratio of the pulses.

By varying the pulse widths throughout a half cycle such that the $\int v \, dt$ value across any interval matches that of a sinewave at the same frequency, a near sinusoidal current can be produced in the load as suggested by Fig. 3.20. As a result, an improvement in performance is obtained through the reduction of the harmonic content in the output waveform. In practice, *m* **chops** of the supply waveform in any quarter cycle of the output waveform can be used to control *m* harmonics, including the fundamental, or to control the modulation index together with $(m - 1)$ harmonics. Emphasis is usually given to the

Arrillaga, J., Bradley, D.A. and Bodger, P.S. (1985). *Power System Harmonics*. John Wiley, U.K.

The modulation index (δ_m) is defined as the ratio of the magnitude of the fundamental component of the output waveform to that of the maximum possible value of the fundamental.

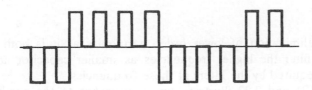

Fig. 3.19 Basic pulse-width-modulated waveform.

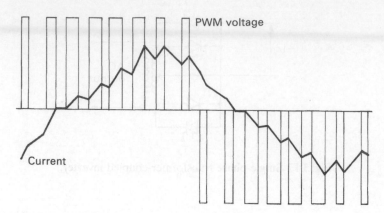

Fig. 3.20 Current waveform for a pulse-width-modulated inverter with an inductive load.

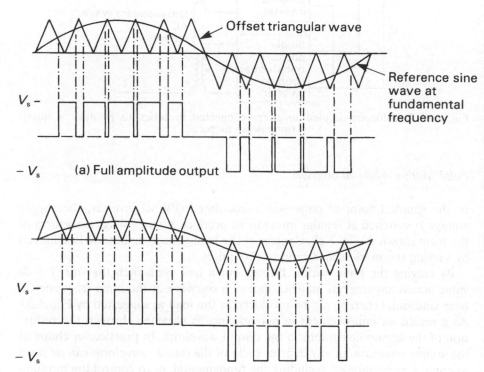

Fig. 3.21 Production of a pulse-width-modulated waveform using an offset triangular wave.

The aim of the PWM control is to match the $\int v \, dt$ value of the output waveform with that of a sinewave at fundamental frequency.

As the switching frequency is increased, then switching loss becomes a major factor in determining inverter performance.

control or elimination of lower order harmonics as it is both easier and cheaper to filter the higher frequencies as smaller capacitor and inductor values are required by the filters at these frequencies.

Figures 3.21 and 3.22 illustrate two approaches to the production of a PWM waveform based on the use of a combination of a reference sinewave at the required frequency and a triangular switching waveform. In each case,

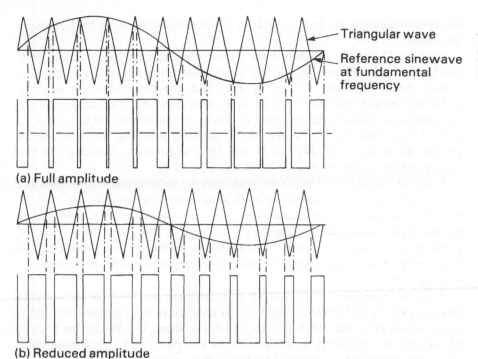

(a) Full amplitude

(b) Reduced amplitude

Fig. 3.22 Production of a pulse-width-modulated waveform using a triangular wave reference.

varying the amplitude of the reference sinewave varies the widths of the individual pulses and controls the effective amplitude of the output waveform.

Implementation is typically by means of special purpose integrated circuits which essentially contain tables of precalculated values of switching angles covering a range of output frequencies for fast access. However, as computational speeds increase, then for the same or reducing cost it may become possible to calculate the required firing angles in real time in order to optimize the strategy for harmonic elimination and control, further improving inverter performance.

Current-sourced inverter

In a current-sourced inverter, the current from the DC source is maintained at an effectively constant level, irrespective of load or inverter conditions. This is achieved by introducing a large inductance in series with the DC supply, enabling changes in the inverter output voltage to be accommodated at low values of di/dt.

The presence of the series inductance also provides protection against the misfiring of individual devices and from short circuits as well as providing a limit on the peak power rating of the individual switching devices. The commutation circuitry of the current-sourced inverter is also simpler than that required by the equivalent voltage-sourced inverter using thyristors and

A current-sourced inverter is typically used to supply high power factor loads whose impedance either remains constant or decreases with increasing frequency.

the resulting circuit can handle reactive loads without the need for freewheeling diodes.

Figure 3.23 shows a single-phase, current-sourced bridge inverter. With thyristors T_1 and T_2 conducting, the commutation capacitors are charged as shown. When thyristors T_3 and T_4 are fired, the capacitors will begin to discharge through thyristors T_1 and T_2 to turn them off. The capacitors will then continue to discharge and then to charge in the reverse direction via the path T_3–D_1–load–D_2–T_4. When the capacitors are fully charged in the reverse direction, diodes D_3 and D_4 will start to conduct, reversing the load current which is now carried by thyristors T_3 and T_4.

A similar commutation sequence is seen in the three-phase current-sourced inverter of Fig. 3.24 in which the individual thyristors conduct for 120°. As with the single-phase current-sourced inverter, commutation is achieved by means of the commutating capacitors C_{13}, C_{35}, etc. With thyristors T_1 and T_2 conducting, capacitor C_{13} will be charged as shown. When thyristor T_3 is fired as the next in sequence, C_{13} initially attempts to discharge through T_1, turning it off.

The diodes have the additional function of isolating the capacitors from any load-voltage transients during commutation.

With an inductive load, current will continue to flow in the original direction via T_3 and diode D_1 until C_{13} is charged in the reverse direction to V_{ba} at which point current begins to transfer to diode D_3. When this transfer of current is completed, commutation is complete and the commutation capacitors appropriately charged for the next commutation sequence. Indeed, T_3 can be turned off by firing either of thyristors T_1 or T_5, enabling any phase sequence to be achieved on the output.

The disadvantages of current-sourced inverters lie primarily with their much slower dynamic response and the need to filter the spikes produced in the output voltage waveform, as illustrated by Fig 3.24b, resulting from the commutation of the individual devices.

Worked example 3.5

A single-phase, current-sourced inverter of the form shown in Fig. 3.23 is being used to supply a load of 20 Ω from a 160 V DC supply at a frequency of 24 Hz. Given that the thyristors have a turn-off time of 50 μs and that the

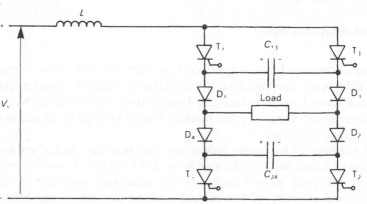

Fig. 3.23 Single-phase current-sourced inverter.

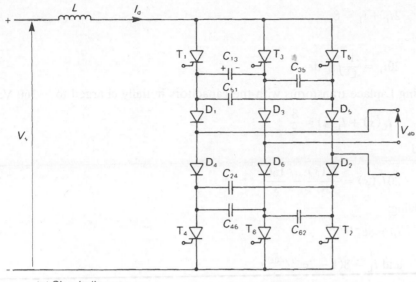

(a) Circuit diagram

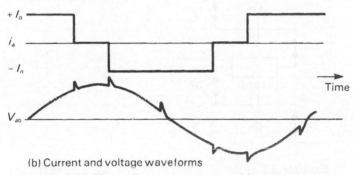

(b) Current and voltage waveforms

Fig. 3.24 Three-phase current-sourced inverter.

maximum permitted value of di/dt under short circuit conditions is 32 A s^{-1}, find suitable values for the series inductance and the commutation capacitors.

The maximum permitted di/dt assumes full voltage and a short circuit on the converter side of the inductance, in which case

$$160 = L\frac{di}{dt}$$

when

$$L = 160/32 = 5 \text{ H}$$

The accompanying figure shows the effective connection of the load immediately following the firing of thyristors T_3 and T_4. The source current will be steady at I_s, where

$$I_s = 160/20 = 8 \text{ A}$$

93

Also

$$2i_C + i_L = 8$$

and

$$20i_L = \frac{1}{C}\int i_C\,dt$$

$I_C(s)$ and $I_L(s)$ are the Laplace transform of i_C and i_L respectively.

Using Laplace transforms with the capacitors initially charged to -160 V:

$$2I_C(s) + I_L(s) = \frac{8}{s}$$

and

$$20I_L(s) = \frac{I_C(s)}{sC} - \frac{160}{s}$$

Solving:

$$i_C = 8e^{-t/40C}$$

and $i_L = 8(1 - 2e^{-t/40C})$

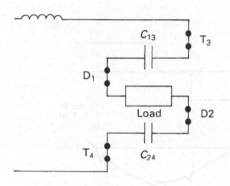

Example 3.5

The reverse voltage must be maintained across thyristors T_1 and T_2 for 50 ms, by which time the load voltage, and hence the load current, will have fallen to zero. Thus

$$e^{-t/40C} = 1/2$$

which gives a value for capacitance of 1.8 μF.

___

Inverter performance

In addition to load-dependent losses in the switching devices a forced-commutated inverter suffers additional losses in the commutation circuit, in protection circuits such as the snubber circuits used to limit dv/dt and as switching losses in the devices. In a PWM inverter the switching losses will be relatively high, reducing overall efficiency and creating heat-removal problems.

Typical efficiencies are 96% for a quasi-squarewave inverter including the

AC/DC converter and DC link, 91% for a PWM inverter using thyristors and 94% for a PWM inverter using transistors or GTO thyristors.

Quasi-squarewave inverters using thyristors normally operate over output frequencies from a few hertz to 100 Hz, or higher in special applications, with transistor inverters operating to 500 Hz. The PWM inverter tends to be limited by switching losses to output frequencies around 100 Hz with operation at higher frequencies in a quasi-squarewave mode.

Current-sourced inverters are typically operated over a frequency range from 5 to 50 Hz, the upper limit being set by the time required for commutation. Commutation losses are, however, small and the overall efficiency to the load is of the order of 96%.

Cycloconverters

A cycloconverter synthesizes its output voltage waveform by switching between the phases of the AC supply and provides an alternative to an inverter where the required range of output frequencies is restricted to less than the supply frequency. The simplest form of cycloconverter is the envelope cycloconverter which produces an output in which each half cycle of the output waveform is made up of a whole number of half cycles of the supply waveform as shown in Figure 3.25. The use of a multi-phase supply enables an output waveform to be produced which approximates to a squarewave as in Fig. 3.25(b). This approach also means that the output is no longer restricted to frequencies which can be made up of integer numbers of half cycles or parts of cycles.

Rissik, H. (1941). *Mercury-arc Current Converter*. Pitman, London.

By controlling the point-on-wave at which the individual phases are switched, an output voltage waveform can be obtained in which the fundamental is emphasized. Figure 3.26 shows the circuit for one phase of a three-pulse cycloconverter from which it can be seen that if the thyristors from the positive and negative groups were fired together then the supply would be short-circuited. To avoid this possibility, **blocked-group** or **inhibited-mode** operation is used with the control circuitry arranged to prevent the simultaneous firing of thyristors in the positive and negative groups.

The waveforms of Fig. 3.26(b) and 3.26(c) illustrate the operation of a three-pulse, blocked-group cycloconverter supplying a resistive and an inductive load respectively. In the latter case, the effect of the load inductance is to shift the phase of the load current with respect to the load voltage, resulting in each group operating in inverting mode for part of each cycle.

By increasing the pulse number of the cycloconverter a better approximation to a sinewave can be obtained in the output voltage.

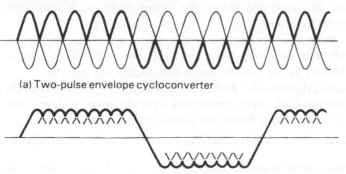

(a) Two-pulse envelope cycloconverter

(b) Six-pulse envelope cycloconverter

Fig. 3.25 Envelope cycloconverter waveforms.

Each group operates
alternately in the rectifying
and inverting modes during
each cycle.

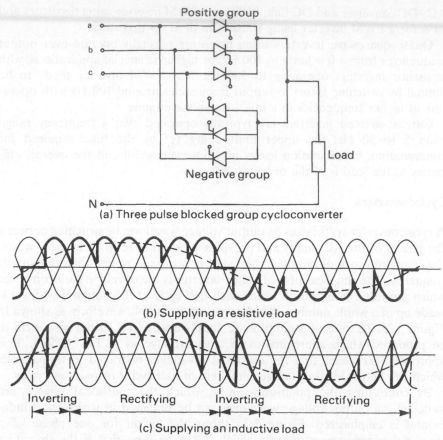

(a) Three pulse blocked group cycloconverter

(b) Supplying a resistive load

Inverting Rectifying Inverting Rectifying

(c) Supplying an inductive load

Fig. 3.26 Operation of a three-pulse blocked-group cycloconverter.

The thyristors are naturally
commutated.

$\hat{V}$ is the peak of the AC
supply voltage.

This is the circulating current
mode of operation.

The peak output voltage ($\hat{V}_o$) that can be provided by a cycloconverter corresponds to the peak DC voltage that each group can supply. Thus for a p-pulse cycloconverter and ignoring overlap:

$$\hat{V}_o = \frac{p}{\pi}\sin(\pi/p)\,\hat{V}\cos\alpha \qquad (3.19)$$

The firing angle of the individual thyristors in the cycloconverter is determined by reference to the instantaneous value of output voltage required. The value of $\cos\alpha$, and hence the firing angle α, is then determined by reference to Equation 3.19.

As an alternative to blocked group operation, a reactor can be connected at the output of the groups as in Fig. 3.27. This allows both groups to conduct simultaneously as the reactor limits any circulating current.

Cycloconverters require more complex control systems than other inverters which tends to limit their application to high power systems with a variable frequency requirement below the supply frequency.

Worked example 3.6

A six-pulse, blocked-group cycloconverter is fed from a three-phase, 600 V (line), 50 Hz supply. The supply has an inductance of 1.146 mH per phase. If

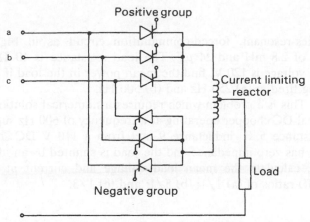

Fig. 3.27 Three-pulse cycloconverter with a current limiting reactor.

the cycloconverter is supplying a variable resistive load with a current of 28 A, estimate the peak and RMS value of load voltage for firing angles of 0°, 30° and 60°.

The peak voltage is the mean voltage of the equivalent rectifier. Equation 2.21 gives

$$V_{\text{mean}} = \frac{p}{\pi} \hat{V}_m \sin\left(\frac{\pi}{p}\right) \cos \alpha - \frac{p \omega L}{2\pi} I_L$$

For this system

$$\frac{p \omega L}{2\pi} I_L = 6 \times 100\pi \times 1.146 \times 10^{-3} \times 28/(2\pi) = 9.63$$

and

$$\frac{p}{\pi} \hat{V}_m \sin\left(\frac{\pi}{p}\right) = 6 \times 660 \times \sqrt{2} \sin(30°) = 891.3$$

(i) $\alpha = 0°$

$$V_{\text{mean}} = 891.3 - 9.63 = 881.7 \text{ V} = V_{\text{max}} \text{ for cycloconverter}$$

Hence the RMS voltage of the cycloconverter is

$$V_{\text{RMS}} = V_{\text{max}}/\sqrt{2} = 623.4 \text{ V}$$

(ii) $\alpha = 30°$

$$V_{\text{mean}} = 891.3 \cos(30°) - 9.63 = 762.3 \text{ V} = V_{\text{max}} \text{ for cycloconverter}$$

Hence the RMS voltage of the cycloconverter is

$$V_{\text{RMS}} = V_{\text{max}}/\sqrt{2} = 539 \text{ V}$$

(iii) $\alpha = 60°$

$$V_{\text{mean}} = 891.3 \cos(60°) - 9.63 = 436 \text{ V} = V_{\text{max}} \text{ for cycloconverter}$$

Hence the RMS voltage of the cycloconverter is

$$V_{\text{RMS}} = V_{\text{max}}/\sqrt{2} = 308.3 \text{ V}$$

3.1 A series-resonant, forced-commutation circuit as in Fig. 3.5(a) has values of 2.8 mH and 24 μF. The load resistance is 40 Ω. If the DC source voltage is 120 V, find the mean power in the load if the thyristor is being fired at (a) 250 Hz and (b) 500 Hz.
[Note: This is a problem which requires a numerical solution.]

3.2 An ideal DC chopper operating at a frequency of 600 Hz supplies a load of resistance 5 Ω, inductance 9 mH from a 110 V DC source. If the source has zero impedance and the load is shunted by an ideal diode as shown, calculate the mean load voltage and current at mark–space (on–off) ratios of (a) 1/1; (b) 5/1; and (c) 1/3.

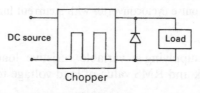

Example 3.6

3.3 A class A chopper of the type shown in Fig. 3.8(a) is operating at a frequency of 2 kHz from a 120 V DC source to supply a load with a resistance of 8 Ω. The load time constant is 6 ms. If the mean load voltage is 72 V, find the required mark–space ratio for the chopper, the mean load current and the magnitude of the current ripple.

3.4 A single-phase bridge inverter such as that of Fig. 3.13 is used to supply a load of 10 Ω resistance, 24 mH inductance from a 360 V DC source. If the inverter is operating at 60 Hz, determine the steady-state power delivered to the load for (a) squarewave operation; (b) quasi-squarewave operation with an 'on' period of 0.6 of a cycle.

3.5 A single-phase, current-sourced inverter such as shown in Fig. 3.22 is used to supply a resistive load of 16 Ω from a 200 V DC source. If the thyristors have a turn-off time of 40 μs and the inverter output frequency is 40 Hz, estimate suitable values for the source inductance and commutating capacitors. Neglect all device voltage drops and losses and assume a maximum di/dt value of 16 A s^{-1}.

3.6 A three-pulse cycloconverter is supplying a single-phase load of 480 V, 72 A at a power factor of 0.85 lagging and a frequency of 25 Hz. Estimate the minimum supply voltage required, the ratings of the thyristors and the power factor of the AC supply. Neglect losses and device voltage drops.

3.7 A three-phase bridge inverter such as that shown in Fig. 3.14 is supplied from a 720 V DC source. For a star-connected load of 18 Ω per phase, find the RMS load current, the load power and the device current ratings for (a) 120° and (b) 180° conduction.

4
Applications I – drives

Objectives

- [ ] To introduce the basic principles of DC and AC induction machines.
- [ ] To consider their operation in the motoring, generating and braking modes.
- [ ] To introduce the principles of variable-speed drives as applied to both DC and AC induction machines.
- [ ] To examine the control of drives based on both machine types.
- [ ] To consider some applications of variable speed drives.
- [ ] To consider briefly other types of electric motor.

Many industrial applications require the provision of a variable-speed rotary drive for their operation. For many years, the employment of an electrical variable speed drive in this role implied the use of a controlled DC machine based on principles established by Harry Ward-Leonard in 1896. Even with the increasing availability of power electronic switching devices, DC machines continued to dominate the variable speed market until the 1980s. From that time on, the growth in availability and rating of self-commutating devices such as GTO thyristors, power transistors and insulated gate bipolar transistors (IGBTs) has meant that drives based on the use of induction machines have become increasingly common.

More recently still, microprocessor-based digital control systems have replaced analogue controllers in many applications, giving an increased sophistication of operation and facilitating the use of other machines such as stepper motors and the switched reluctance motor.

DC machines

A conventional DC machine consists of a stationary field winding and a rotating armature winding as shown in Fig. 4.1. The field winding is supplied with a DC current to produce an essentially static magnetic field distribution within the airgap of the machine. This magnetic field then interacts with the current in the armature conductors to produce a torque to rotate the armature. In order to sustain this torque, the armature current distribution relative to the field must be maintained constant, irrespective of the actual rotor position. This is achieved by the action of the commutator which acts to reverse the direction of the current in the armature conductors as they pass from under one field pole to the next.

Also, as the armature conductors are moving through the magnetic field produced by the field winding they have a voltage, or back EMF (E_a) induced in them which is a function of the strength of the field and the speed of

Ward-Leonard. H., (1896). Volts versus Ohms – The speed regulation of electric motors. *AIEE Transactions*, **13**, 375–84.

Ward-Leonard drive; a DC generator is used to provide a variable voltage supply for a DC motor.

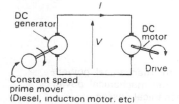

Bradley, D.A. (1994). *Basic Electric Power and Machines*. Chapman & Hall, London.

The force on a conductor in a magnetic field is given by

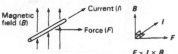

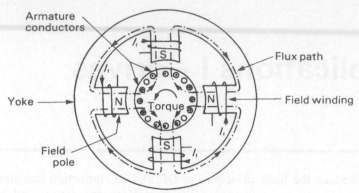

Fig. 4.1 Four-pole DC machine.

The voltage induced in a conductor length L metres, moving with a velocity $v\,\mathrm{m\,s^{-1}}$ in a magnetic field of strength B tesla is

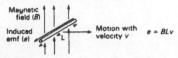

rotation of the armature. This back EMF will oppose the applied voltage on the armature when the DC machine is acting as a motor and provide the source voltage when it is used as a generator.

When motoring, the DC machine draws power from the DC source and the electrical torque developed by the machine then acts to rotate the armature against the mechanical load applied to the armature. In generating mode, power is delivered by the machine to the DC load and the electrical torque developed opposed the applied mechanical torque driving the armature. The defining steady-state equations are, with reference to Fig. 4.2:

The general equation for the armature has the form:

$$E_a = V \pm I_a R_a$$

in which + represents motoring and − represents generating.

$$\text{Motoring} \quad E_a = V + I_a R_a \tag{4.1}$$

$$\text{Generating} \quad E_a = V - I_a R_a \tag{4.2}$$

and

$$E_a = K\phi\omega \tag{4.3}$$

where ϕ is the flux per pole
and ω is the rotational speed of the armature in radians per second.

The total mechanical power developed by the DC machine, including mechanical losses is

Useful mechanical power = Internal mechanical power − Mechanical losses.

$$\text{Internal mechanical power} = T\omega = E_a I_a \tag{4.4}$$

The field winding of a DC machine can be supplied by a shunt or a series field winding or some combination of these:

Shunt field

In a shunt-connected DC machine the field current is derived from the same supply as the armature as shown in Fig. 4.3(a). The field current is,

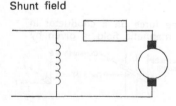

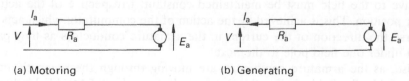

(a) Motoring
(b) Generating

Fig. 4.2 DC machine armature circuit.

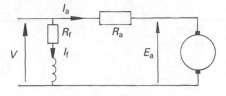

(a) DC shunt connected motor

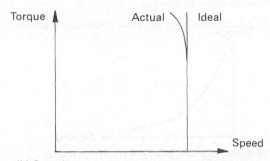

(b) Speed/torque characteristic of a DC shunt motor

Fig. 4.3 DC shunt motor connections and characteristics.

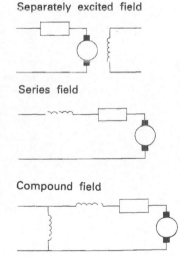

Separately excited field

Series field

Compound field

however, independent of the armature current. The resulting torque–speed characteristic for this connection is then as shown in Fig. 4.3(b).

In a separately excited DC machine the field current is derived from its own independent supply as in Fig. 4.4. The resulting torque–speed characteristic is the same as in Fig. 4.3(b) for the shunt-connected DC machine.

In a series-connected DC machine, all or part of the armature current flows through the field winding as in Fig. 4.5(a). The proportion of the armature current that goes to the field can be controlled by the diverter resistance included in this figure. The resulting torque–speed characteristic is then as shown in Fig. 4.5(b).

A compound wound DC machine contains both series and shunt (or separately excited) field windings as illustrated in Fig. 4.6. The resulting field can either be the sum (cumulative compounding) or the difference (differential compounding) of the applied series and shunt fields.

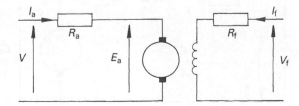

Fig. 4.4 Separately excited DC motor.

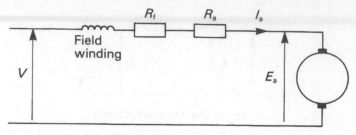

(a) DC series connected motor

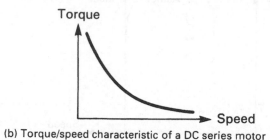

(b) Torque/speed characteristic of a DC series motor

Fig. 4.5 DC series motor connections and characteristics.

Open circuit characteristic

The open circuit characteristic or excitation curve of a DC machine has the general form Fig. 4.7 and is obtained by driving the armature of the machine at constant speed and recording the open circuit voltage (E_a) over the full range of field current. The shape of the curve is determined by the magnetic characteristics of the machine. Important features of the open circuit characteristic are:

The production of a small voltage with zero applied field. This voltage results from the residual or remnant flux remaining in the core of the machine once the main field is removed.

The open circuit voltage is not a linear function of the field current. This occurs since the machine is constructed from ferromagnetic materials which saturate at higher values of field, causing the roll over of the curve at the higher values of field current.

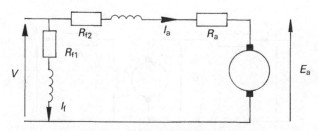

Fig. 4.6 Compound DC motor.

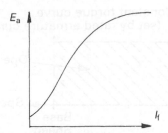

Fig. 4.7 Open circuit characteristic of a DC machine.

Only a single curve is needed to describe the operation of the DC machine over the whole of its speed range since, for constant field, the relationship

$$\frac{E_{a0}}{E_{a1}} = \frac{\omega_0}{\omega_1} = \frac{n_0}{n_2} \tag{4.5}$$

ω_0 and ω_1 are machine speeds measured in radians/s while n_0 and n_1 are machine speeds measured in revolutions per minute.

applies, enabling the open circuit voltage or back EMF to be calculated at any speed.

Motoring

Neglecting armature resistance then, from Equations 4.3 and 4.4:

$$V = E_a = K\phi\omega \tag{4.6}$$

If the applied armature voltage V is held constant, it can be seen from Equation 4.6 that the motor speed ω can be controlled by varying the field ϕ such that

Field control.

$$\omega \propto 1/\phi \tag{4.7}$$

Under these conditions and setting the limiting condition for continuous operation as the maximum continuous current that can be carried by the armature winding of the machine, then, using, Equations 4.4 and 4.6 with the applied armature voltage V, and hence E_a, constant, the DC machine is seen to operate with a constant power limit. This is shown in Fig. 4.8.

The maximum continuous armature current is referred to as the rated current of the DC machine.

For constant power, torque $\propto$ $1/\omega$.

The minimum speed obtainable with field control is determined by the maximum armature voltage and the maximum field current and is the base

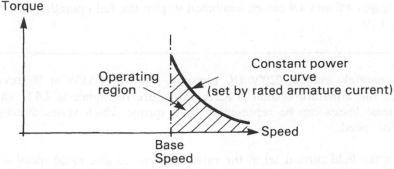

Fig. 4.8 Operating region of a DC motor with field control.

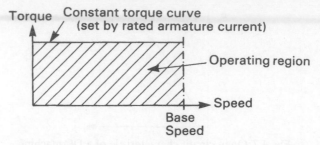

Fig. 4.9 Operating region of a DC motor with armature voltage control.

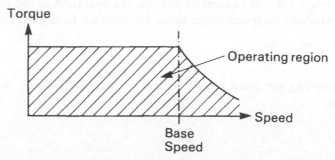

Fig. 4.10 Operating region combining field and armature voltage control.

speed of Fig. 4.8. The upper limit of speed is set largely by the commutation behaviour of the machine under weak field conditions. Speed variation of the order of 4 or 5 to 1 is typically obtainable for small machines, falling to around 2 to 1 for large machines.

Referring again to Equation 4.6, if the field ϕ is held constant then the speed of the motor can be controlled by varying the voltage applied to the machine armature when

$$V \propto \omega \tag{4.8}$$

Applying the same restriction on the maximum value of armature current as before, then, by reference to Equation 4.4, the DC motor is found to operate with a constant torque limit as in Fig. 4.9. The maximum operating speed is achieved with maximum applied armature voltage and maximum field current with an operating speed range of around 100 to 1.

Figures 4.8 and 4.9 can be combined to give the full operating envelope of Fig. 4.10.

Worked example 4.1

A separately excited, 220 V DC motor is rated at 1.5 kW at 900 rev min^{-1} when the armature current is 8.2 A. Armature resistance is 2.4 Ω and mechanical losses can be represented by a torque which varies directly with motor speed.

With the field current set at the value required to give rated speed at rated load and armature voltage, find the values of armature voltage required and

104

the armature current when the motor is operating against a load of 3.5 N m at rated speed. Saturation may be neglected.

To what value would the armature voltage have to be adjusted to drive the same 3.5 N m load at 750 rev min^{-1}?

For operation at rated speed and load:

Input power to motor $= VI_a = 220 \times 8.2 = 1804$ W

$I_a^2 R_a$ loss $= 8.2^2 \times 2.4 = 161.4$ W

Mechanical losses = Input power − Output power − $I_a^2 R_a$ loss

$\qquad\qquad\qquad = 1804 - 1500 - 161.4 = 142.6$ W

Mechanical loss torque $= \dfrac{142.6}{2\pi (900/60)} = 1.51$ N m

Now

$\qquad E_a = V - I_a R_a = 220 - 8.2 \times 2.4 = 200.3$ V

Using $E_a = K\phi\omega$ gives

$$K\phi = \frac{200.3}{2\pi (900/60)} = 2.125$$

When operating against the 3.5 N m load at rated speed, the total torque developed by the motor is

$\qquad$ Motor torque = 3.5 + Mechanical loss torque = 5.01 N m

Now, torque $T = K\phi I_a$ when

$$I_a = \frac{T}{K\phi} = \frac{5.01}{2.125} = 2.36 \text{ A}$$

and

$\qquad V = E_a + I_a R_a = 200.3 + 2.36 \times 2.4 = 205.9$ V

At 750 rev min^{-1}

$\qquad$ Mechanical loss torque $= 1.51 \times \dfrac{750}{900} = 1.26$ N m

Then

$\qquad$ Motor torque = 3.5 + 1.26 = 4.76 N m

when

$$I_a = \frac{4.76}{2.125} = 2.24 \text{ A}$$

and

$$E_a = K\phi\omega = 2.125 \times 2\pi \times \frac{750}{60} = 166.9 \text{ V}$$

when

$\qquad V = 166.9 + 2.24 \times 2.4 = 172.3$ V

Braking

During variable-speed operations the DC motor will be required both to accelerate and decelerate. The deceleration process can be assisted where required by one or other of the various forms of electrical braking shown in Fig. 4.11.

In Fig. 4.11(a), the armature is disconnected from the supply and connected to a braking resistor. With a field applied to the machine, an EMF is produced in the armature winding which drives a current through the braking resistor, dissipating the energy stored in the rotating inertia of the machine armature.

By reversing the connection of the armature to the DC supply, as in Fig. 4.11(b), the back EMF will add to the supply voltage and drive a reverse current through the armature of the machine. This produces a reverse torque which acts to decelerate the motor. This is a very severe operating condition for the DC machine and a current limiting resistor would normally be included in series with the armature to control the armature current.

If the field of the DC machine is adjusted so that the back EMF is greater than the voltage applied to the armature ($E_a > V$) the machine will act as a generator transferring energy from the mechanical system to the DC supply as in Fig. 4.11(c). The maximum allowable field is limited by the field voltage supply and the maximum permitted field current.

DC machine dynamics

Consideration of the operation of a DC machine has so far been in terms of its steady-state performance. In practice, applications such as robots involve the dynamic behaviour of the machine. Such situations require the effects of the machine armature and field inductances, the rotary inertia and the non-linearities of the magnetic circuit to be taken into account.

For dynamic analysis the machine equations are rewritten using instanta-

> Resistive or dynamic braking.

> Reverse current braking or 'plugging'.

> The armature voltage must be removed once the speed has been reduced to zero to prevent the direction of rotation from being reversed.

> Regenerative braking.

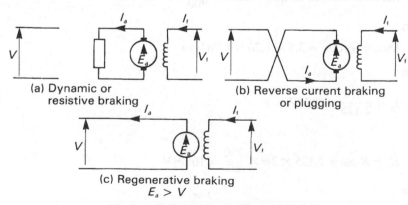

(a) Dynamic or resistive braking

(b) Reverse current braking or plugging

(c) Regenerative braking
$E_a > V$

Fig. 4.11 DC machine braking modes.

neous values, in which case

$$v = e_a \pm (i_a R_a + L_a \, di_a/dt) \qquad (4.9)$$

$$T = K\phi i_a = T_L + J \, d\omega/dt \qquad (4.10)$$

where T_L is the load torque
J is the rotary inertia
and L_a is the armature inductance.

Also

$$e_a = K\phi\omega \qquad (4.11)$$

For the field circuit

$$v_f = i_f R_f + L_f \, di_f/dt \qquad (4.12)$$

and

$$\phi = f(i_f) \qquad (4.13)$$

where L_f is the inductance of the field circuit.

A separately excited DC machine is used to drive a load whose torque is directly proportional to the speed of the machine. Sketch the block diagram for this system and hence derive the transfer function between speed and armature voltage with field current held constant.

The armature of the machine has a resistance of 1.2 Ω and an inductance of 0.08 H and the machine has a torque constant of 1.82 N m A^{-1}. The combined inertia of the load and machine armature is 12 kg m^2 and the load torque constant is 0.06 N m s rad^{-1}.

Find the final speed of the machine when a supply of 48 V is suddenly applied to the armature.

Worked example 4.2

The block diagram for the system is as shown, in which case

$$\omega(s) = \frac{K(V(s) - K\omega(s))}{Js(R_a + L_a s)} - \frac{K_T \omega(s)}{Js}$$

Solving for $\omega(s)$ gives

$$\omega(s) = \frac{KV(s)}{Js(R_a + L_a s) + K^2 + K_T(R_a + L_a s)}$$

Substituting values

$$\omega(s) = \frac{1.82}{12s(1.2 + 0.08s) + 3.312 + 0.06(1.2 + 0.08s)} \frac{48}{s}$$

$$= \frac{1.82}{0.96s^2 + 14.4048s + 4.032} \frac{48}{s}$$

$V(s)$, $I_a(s)$, $T_L(s)$ and $u(s)$ are the Laplace transforms of the machine parameters.

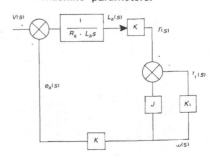

Taking partial fractions:

$$\omega(s) = \frac{21.67}{s} - \frac{21.67(s + 7.503)}{(s + 7.503)^2 - 52.09} + \frac{161.15}{(s + 7.503)^2 - 52.09}$$

Taking inverse Laplace transforms gives

$$\omega = 21.67 - 21.67e^{-7.503t}\cosh(7.22t) + 22.32e^{-7.503t}\sinh 7.22t \text{ rads s}^{-1}$$

The steady-state speed is then

$$\omega = 21.67 \text{ rads s}^{-1} \text{ (or } 206.9 \text{ rev min}^{-1})$$

Variable-speed DC drives

The most common form of DC variable-speed drive is that based on the control of armature voltage using a fully-controlled or half-controlled converter as in Fig. 4.12(a). The majority of converter-based drives are intended for operation with a DC machine having sufficient armature inductance to maintain the armature current sensibly constant over one cycle of the fundamental voltage waveform. In applications where matching is less critical, then appropriate inductance may be added in series with the armature of the machine.

An alternative, shown in Fig. 4.12(b), would be to use a combination of a controlled or uncontrolled converter together with a DC chopper. A DC chopper would also be used where a DC source was already available.

If regeneration is required a fully-controlled converter must be used.

DC machines intended specifically for variable speed operation from a controlled converter are designed with laminated field poles and increased armature inductance.

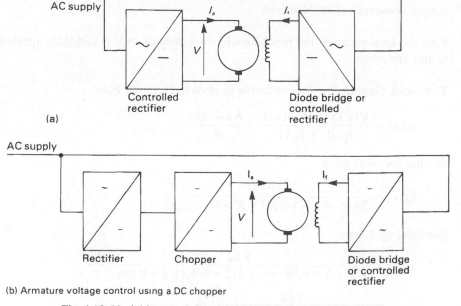

(b) Armature voltage control using a DC chopper

Fig. 4.12 Variable speed drives using armature voltage control.

For lower rated drives using half-controlled converters with low levels of armature circuit inductance, a commutating diode is included as in Fig. 4.13. As has been demonstrated, this prevents any reversal of the ouput voltage of the converter and results in improved commutation and reduced armature current ripple. Under conditions of light load or on starting the armature current may however become discontinuous.

Freewheeling or bypass diode (Chapter 2).

Reversing drives

The direction of rotation of a DC motor can be reversed by reversing either the armature voltage or the field current. Contactor reversal as shown in Fig. 4.14(a) can be used with either a fully-controlled or a half-controlled converter as the source of armature voltage, with regenerative braking being possible in the former case. In operation, the speed of the machine in the original direction and the armature current are first reduced to zero after which the contactor can be operated to reverse the direction of the applied armature voltage.

Contractor reversal.

The use of the contactor introduces a period of around 0.2s over which no torque is developed. While this may be acceptable for drives such as presses, hoists, lathes and marine propulsion where reversals are infrequent, there are applications, such as steel strip mills and paper mills, where a more rapid reversal is required.

Full four-quadrant operation can be achieved by using a pair of bridges as in Fig. 4.14(b). Assuming initial operation in motoring mode with bridge B_1 conducting then reversal would take place as follows:

Dual bridge reversing drives.

The firing angle of bridge B_1 is increased to turn-off the current in the bridge.
Bridge B_1 is turned off.
Bridge B_2 takes over in inverting mode and the machine regenerates, returning energy to the DC supply.
As the speed of the machine reduces, the firing angle of bridge B_2 is reduced to give zero voltage at or near to zero speed.
The firing angle of bridge B_2 is further reduced to control the acceleration of the machine in the reverse direction.

Half-controlled converter

Fig. 4.13 DC machine control using a half-controlled converter with a commutating or freewheeling diode.

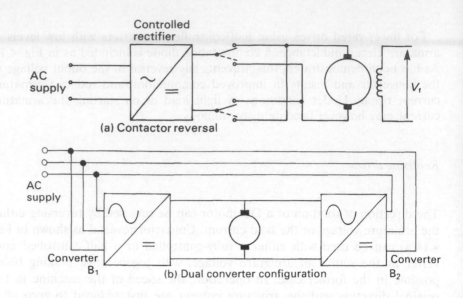

(a) Contactor reversal

(b) Dual converter configuration

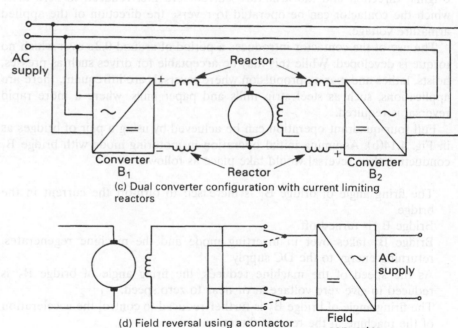

(c) Dual converter configuration with current limiting reactors

(d) Field reversal using a contactor

Fig. 4.14 Reversing drives.

The machine is now running as a motor in the reverse direction.

A further improvement in performance can be achieved by including current limiting reactors between the forward and reverse bridges as in Fig. 4.14(c). The two bridges can now be allowed to conduct simultaneously, reducing the time required for reversal.

Field reversal can also be achieved by means of either contactors as in Fig. 4.14(d) or dual bridges. The reversal of the field current can only take place

Field reversal.

relatively slowly because of the need to remove the stored energy in the field prior to current reversal, introducing delays of about 1 s into torque reversal.

Field forcing, in which a high initial voltage is applied to the field circuit, is used to produce a rapid initial rise of field current in order to speed up torque reversal. The field voltage is then reduced to the value required to maintain the desired operating field current.

Control

A drive control system may be required to perform a combination of functions including:

(a) To respond to changes in demand speed or torque.
(b) To provide start-up and shut-down procedures.
(c) To ensure that operation is largely independent of fluctuations in supply conditions.
(d) To optimize the operating conditions for best performance.

The control system may also be required to provide protection against overloads and faults, to maintain a check on drive status and performance, to synchronize the operation of a number of drives, to provide field forcing and operate reversing drives.

A basic control system for a variable-speed DC drive is shown in Fig. 4.16. Information on output speed or position is fed back to the comparator, the output of which is used via the error amplifier to control the firing circuits for the converter. Because of the low armature resistance of the DC motor, this simple arrangement could result in excessive currents being drawn from the converter, particularly where large changes in load or speed are involved and some means of limiting armature current is therefore usually included in the control circuit.

Figure 4.15 shows the machine operating modes during reversal.

The converter in this case is acting as a power amplifier.

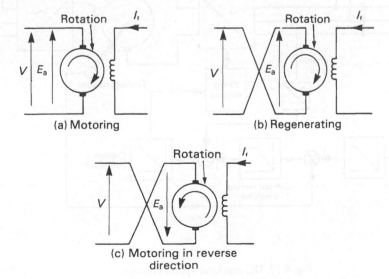

(a) Motoring

(b) Regenerating

(c) Motoring in reverse direction

Fig. 4.15 Operating modes during reversal.

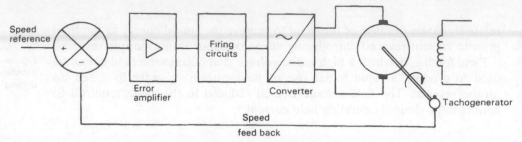

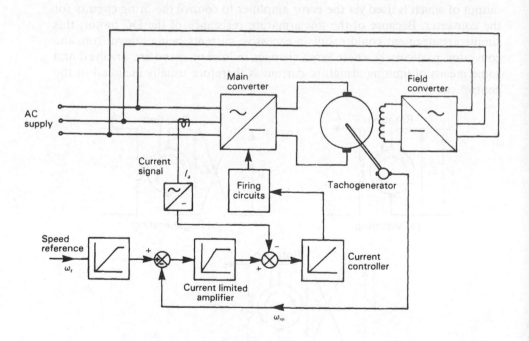

Fig. 4.16 Basic control system.

Figure 4.17 shows a typical analogue controller for a single quadrant drive. The demand or reference speed (ω_r) is fed to the comparator via a ramp generator to smooth out and limit the effect of sudden or large changes in demand speed. At the comparator, the reference is compared with the actual speed (ω_m) obtained from a tachogenerator on the shaft of the DC machine. The resulting comparator error signal is then taken to a current limited amplifier which restricts the maximum error signal and hence the maximum armature current. The output of the current limited amplifier is then compared with the armature current (I_a) of the DC motor which may be obtained either by means of a direct current current transformer (DCCT) or a Hall effect sensor in the armature circuit or an AC transformer and rectifier in the converter supply as in the figure. The output of this second comparator is then used to control the firing angle of the converter.

The direct current current transformer is described briefly in Chapter 6 (Fig. 6.3 and associated text).

Fig. 4.17 DC machine speed control.

In the form shown in Fig. 4.17 and described above, the motor will operate with a steady-state speed error. This could be eliminated by modifying the controller to incorporate a full Proportional, Integral and Derivative (PID) or Three-Term control scheme.

The speed error arises from the requirement that there must be an error signal at the first comparator to provide a signal to control the firing circuits. The actual error will be a function of the speed of the machine and the system gain.

Analogue controllers of the type described can provide a speed stability of about 0.1%. Digital control systems based on the microprocessors and, more recently, on application specific integrated circuits (ASICs) have been developed. These offer control systems of greater precision, flexibility, consistency, stability and noise immunity with a speed stability of around 0.01% and have largely replaced analogue systems for many applications. In addition, digital systems enable the precise speed matching or controlled speed ratios between two or more motors by the use of a common speed reference and could also enable the phase relationship between the shafts of individual motors to be controlled.

Electronic gearbox.

The development of ASIC-based control implementations when combined with hybrid thick film technologies also offers the opportunity for embedding the complete controller within the frame or enclosure of smaller DC machines, simplifying the overall system requirements.

Case study – servo amplifier

Because of their flexibility and ease of control, DC machines are often used as servomotors in a range of applications ranging from robots to positioning systems and are capable of providing a wide range of motion profiles.

DC servomotors are specially designed to enable rapid response and flexibility.

Typically, a DC servomotor will have a characteristic similar to that of a shunt or separately excited machine leading to the basic configuration of a typical servomotor drive shown in Fig. 4.18. The servo amplifier, the specification for which is given in Table 4.1, is based on the circuit of Fig. 4.19 and

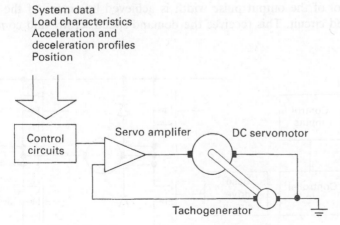

Fig. 4.18 DC servomotor for a single-axis drive.

Table 4.1 Servo Amplifier Specification

Input		*Output*	
Power supply	75–90 V DC at 7.5 A continuous 20 A peak	Continuous current	7.5 A
Velocity	± 10 V DC	Peak current	20 A for 5 s
Tacho	12–20 V DC at maximum motor speed	Deadband	Zero deadband
		Output waveform	PWM at 3 kHz
		Indicator lamps	Power on
Current limit	Adjustable from 10 to 20 A		Forward enabled
			Reverse enabled
Tacho gain	Adjustable between 12–10 V for maximum motor speed		Overtemperature trip
			Overcurrent trip
			Overvoltage trip
			Current limit
Zero offset	Adjustable to ± 4% of velocity input	Fault relay	Volt-free contact rated at 240 V AC or 200 V DC at 0.5 A closes on operation of any of the three trips
Forward enable	Enables forward rotation when connected to 0 V		
Reverse enable	Enables reverse rotation when connected to 0 V		
Drive enable	Enables forward and reverse rotation when connected to 0 V		
General			
Ambient temperature	0–40°C	Dust and moisture	IP 22

is configured to provide full, four-quadrant operation of the servomotor using a pulse width modulated chopper operating at a frequency of 3 kHz.

Control of the output pulse width is achieved by means of the controller integrated circuit. This receives the demand speed setting and compares this

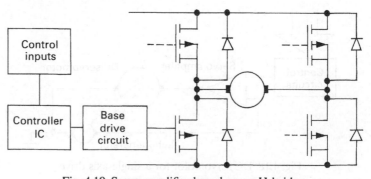

Fig. 4.19 Servo amplifier based on an H-bridge.

114

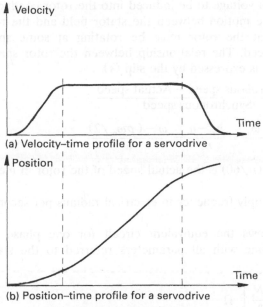

(a) Velocity–time profile for a servodrive

(b) Position–time profile for a servodrive

Fig. 4.20 Servomotor operation.

with actual speed information supplied by the tachometer. Position data could also be provided by an appropriate encoder on the machine shaft if desired.

The controller integrated circuit then adjusts the firing angles of the power MOSFETS to balance the demand and actual speeds. The demand speed is set by the external system which also defines the required acceleration and deceleration profiles for the servomotor. A typical operating profile when used for positioning would be as shown in Fig. 4.20(a), with the resulting motion profile as shown in Fig. 4.20(b).

The servo amplifier incorporates within itself the necessary directional control logic, the condition of which is set by the external system, together with appropriate status and alarm logic. Inbuilt trips include both current and voltage limits as well as temperature.

Power MOSFETS are described in Chapter 1.

AC machines

The most commonly used form of AC machine is the polyphase induction machine. In the polyphase induction machine, the stator winding is supplied with a balanced set of currents which set up a magnetic field rotating at synchronous speed in the airgap of the machine. This rotating magnetic field then interacts with the rotor conductors, inducing a voltage, and hence a current, into these conductors. The current in the rotor conductors then interacts with the rotating magnetic field to produce a torque which acts to rotate the rotor in the same direction as the rotating magnetic field produced by the stator winding.

Synchronous speed (rev min^{-1}) $= n_s = 120f/p$ where f is the supply frequency and p the number of poles on the machine. Mechanical synchronous speed in mechanical radians per second (rad s^{-1}) $= \omega_{sm} = 2\pi(n_s/60) = 2\pi f \times 2/p = 2\omega/p$ where ω is the angular frequency of the supply in electrical radians per second (rad s^{-1}).

115

In order for a voltage to be induced into the rotor conductors, there must be some relative motion between the stator field and the rotor conductors. This means that the rotor must be rotating at some speed other than synchronous speed. The relationship between the rotor speed and the synchronous speed is expressed by the slip (s).

$$s = \frac{\text{Synchronous speed} - \text{Actual speed}}{\text{Synchronous speed}}$$

$$= \frac{\omega_{sm} - \omega_m}{\omega_{sm}} = \frac{n_s - n}{n_s} = \frac{\omega - (p\omega_m/2)}{\omega} \qquad (4.14)$$

where $\omega_m = 2\pi(n/60)$ is the actual speed of the rotor in mechanical radians per second
and ω is the supply frequency in electrical radians per second

Figure 4.21 shows the equivalent circuit for one phase of a polyphase induction machine with all parameters referred to the stator winding, in which case

$$R_r' = R_r \left[\frac{N_s}{N_r} \right]^2 \ \Omega \qquad (4.15)$$

and

$$X_r' = X_r \left[\frac{N_s}{N_r} \right]^2 \ \Omega \qquad (4.16)$$

where R_r is the resistance of the rotor winding (Ω)
X_r is the reactance of the rotor winding (Ω)
and N_s/N_r is the stator/rotor turns ratio for the machine

The equivalent circuit of Fig. 4.21 can be simplified by combining the stator and magnetizing branch components to give the Thevenin equivalent circuit of Fig. 4.22.

The total mechanical power developed by one phase of the machine is P_m when

$$P_m = T_m \omega_m \qquad (4.17)$$

where T_m is the equivalent torque
and ω_m is the speed of the rotor in mechanical radians per second.

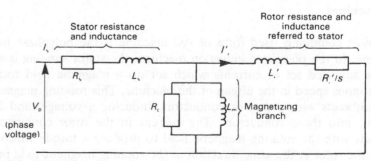

Fig. 4.21 Equivalent circuit of one phase of an induction machine.

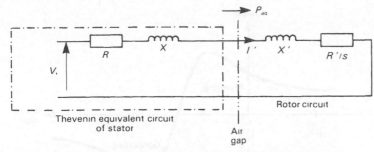

Fig. 4.22 Thevenin equivalent circuit of an induction motor.

The power transferred across the airgap of the machine to the rotor is, in terms of the equivalent circuit:

$$P_{ag} = I_r'^2 R_r'/s \qquad (4.18)$$

The power lost in the rotor circuit is

$$P_r = I_r'^2 R_r' \qquad (4.19)$$

The power converted to mechanical power (P_m) is therefore

$$P_m = P_{ag} - P_r = I_r'^2 R_r'(1-s)/s = P_{ag}(1-s) \qquad (4.20)$$

Now, from Equation 4.14

$$\omega_m = (1-s)\omega_{sm} \qquad (4.21)$$

In which case

$$T_m = P_m/\omega_m = P_{ag}/\omega_{sm} = P_{ag} \times p/(2\omega)\,\text{N m} \qquad (4.22)$$

For a three-phase machine, the total torque (T_m) may be expressed in terms of the airgap power in one phase (P_{ag}) as

$$T_m = 3P_{ag}\,p/(2\omega)$$

$$= \frac{3p}{2\omega}\left[\frac{V_t^2}{(R+R_r'/s)^2 + (X+X_r')^2}\right]\frac{R_r'}{s}\,\text{N m} \qquad (4.23)$$

The generalized curve of torque against slip (or speed) can now be obtained. A typical form is shown in Fig. 4.23 and can be seen to contain three regions as follows.

In the motoring region of the torque–slip characteristic, the slip has a value between 1 and 0 and the machine is developing a mechanical output torque. Operation is typically with low values of slip on the linear portion of the characteristic near to synchronous speed giving near constant speed operation for a constant voltage and frequency supply.

In the generating region of the characteristic, slip is negative and the electrical torque produced by the machine opposes the applied mechanical torque and energy is returned to the AC supply.

The slip at which maximum torque is obtained is given by

$$s_{max} = R_r'/[R^2 + (X+X_r')^2]^{1/2}$$

and the maximum torque is then

$$T_{max} = \frac{3p}{4\omega}\left[\frac{V_t^2}{R+[R^2+(X+X_r')^2]^{1/2}}\right]$$

Motoring.

Generating.

117

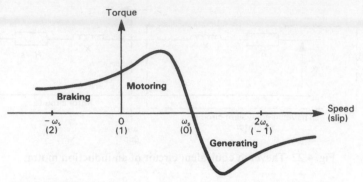

Fig. 4.23 Torque–speed characteristic of an induction machine.

Plugging.

In the braking region of the characteristic, slip is greater than 1 and the torque developed opposes the direction of rotation of the machine. This is a very severe condition involving very high stator currents.

Worked example 4.3

A 50 Hz, three-phase, 415 V (line), four-pole, slip-ring induction motor when running at a speed of 1415 rev min^{-1} has windage and friction losses of 520 W. The machine Thevenin equivalent circuit parameters per phase are:

$R_r' = 0.113\,\Omega$
$X_r' = 0.405\,\Omega$
$R = 0.239\,\Omega$
$X = 0.398\,\Omega$
$V_t = 238.6\,V$

Estimate the mechanical torque developed, the useful mechanical power and the efficiency of the machine.

Synchronous speed $= 120f/p = (120 \times 50)/4 = 1500$ rev min^{-1}

When

Slip $= (1500 - 1415)/1500 = 0.05667$

In which case

$R_r'/s = 0.113/0.05667 = 1.994\,\Omega$

Using Equation 4.22,

$$T_m = \frac{3 \times 4}{2 \times 314.16}\left[\frac{238.6^2}{(1.994 + 0.239)^2 + (0.405 + 0.398)^2}\right] 1.994 = 385\,\text{N m}$$

The useful mechanical power is then

$$P_o = 385 \times 2\pi \times (1415/60) - 520 = 56\,530\,\text{W}$$

The input power and hence the efficiency can be estimated using the Thevenin equivalent circuit when, with V_t as reference:

$$I_r' = 238.6(2.233 + 0.803\text{j}) = 100.55\,|-19.78°$$

The input power is then

$$P_{in} = 3 \times 238.6 \times 100.55 \times \cos(19.78°) = 67\,727\,\text{W}$$

Hence, efficiency

$$\eta = 100 \times (56\,530/67\,727) = 83.47\%$$

Induction machine operation and control

If the magnitude of the AC voltage applied to the induction machine is varied at a fixed frequency, then, by reference to the expression for maximum torque, the resulting family of torque–speed characteristics will be as shown in Fig. 4.24. This will enable a limited degree of speed control but at the cost of a significant decrease in efficiency with decreasing speed.

Voltage control.

A simple means of achieving voltage control is to use a reverse parallel pair of thyristors in each phase of the supply as in Fig. 4.25; its use is, however, accompanied by an increase in the level of the harmonic currents drawn from the supply.

This arrangement is also used to provide a reduced voltage to the machine during acceleration and run-up. Typically, upper and lower limits will be set for the line currents during run-up and the controller will then control the firing angle of the thyristors to maintain the line current within these bounds until the point at which conduction becomes continuous. The firing angle is then reduced to zero and the machine operates normally.

Soft start.

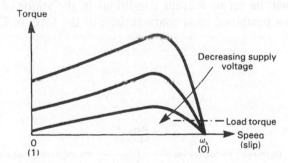

Fig. 4.24 Torque–speed characteristic with voltage control.

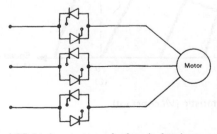

Fig. 4.25 Voltage control of an induction motor.

The maximum torque of the induction machine is given by Equation 4.24:

$$T_{max} = \frac{3p}{4\omega}\left[\frac{V_t^2}{R + \left[R^2 + (X + X_r')^2\right]^{1/2}}\right] \tag{4.24}$$

Ignoring resistance R, Equation 4.24 may be approximated by

$$T_{max} = \frac{3p}{4\omega}\frac{V_t^2}{(X + X_r')^2} = \frac{3p}{4(L + L_r')}\frac{V_t^2}{\omega^2} \tag{4.25}$$

Therefore, for constant T_{max}:

$$V_t/f = \text{constant} = V/f \tag{4.26}$$

The use of an ideal machine together with variable frequency source with a constant V/f ratio would result in a family of characteristics such as that shown in Fig. 4.26(a).

In practice, the effect of stator resistance in particular will be to modify the characteristic envelope as shown in Fig. 4.26(b) which shows a significant fall-off in capability at low speed. In order to overcome this effect at least partially, the voltage applied to the machine would normally be increased at low speeds beyond the ideal value suggested by Equation 4.26, as in Fig. 4.26(c).

If operation at frequencies above the normal supply frequency is required then this would typically be at constant voltage, in which case, reference to equation 4.25 suggests a power limited characteristic of the form shown in Fig. 4.27.

Compare Fig. 4.28 with Fig. 4.10 for a DC machine.

The actual limit to machine torque will not be the maximum torque available but will be set by current conditions in the stator of the machine. This results in a combined ideal characteristic of the form of Fig. 4.28.

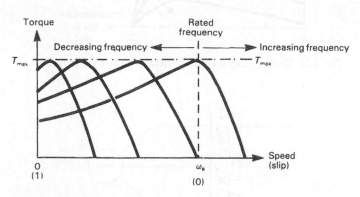

(a) Ideal characteristic (V/f constant)

Fig. 4.26 Voltage and frequency control of an induction motor.

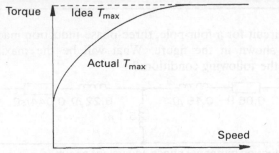

(b) Characteristic showing low speed affects (V/f constant)

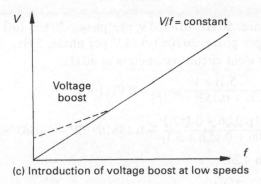

(c) Introduction of voltage boost at low speeds

Fig. 4.26 *contd*

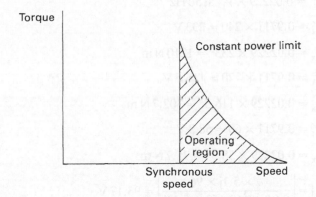

Fig. 4.27 Operation of an induction machine with constant voltage and variable frequency.

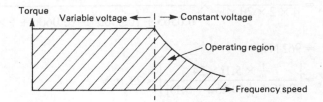

Fig. 4.28 Operating envelope of an induction motor with variable frequency control.

The equivalent circuit for a four-pole, three-phase induction machine operating at 50 Hz is shown in the figure. What will be the maximum torque produced under the following conditions?

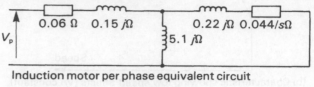

Induction motor per phase equivalent circuit

Example 4.1

(i) 240 V per phase, 50 Hz; (ii) 120 V per phase, 50 Hz; (iii) 60 V per phase, 50 Hz; (iv) 96 V per phase, 20 Hz; (v) 24 V per phase, 5 Hz.

Thevenin equivalent circuit parameters at 50 Hz:

$$|V_t| = \left| \frac{5.1j \times V}{0.06 + 0.152j + 5.1j} \right| = 0.9711 \times V$$

$$Z_t = \frac{5.1j(0.06 + 0.152j)}{0.06 + 0.152j + 5.1j} = 0.158 | 69.12° = 0.0563 + 0.148j \; \Omega$$

From Equation 4.24,

$$T_{max} = \frac{3 \times 4}{4 \times 2 \times 50 \times \pi} \times \frac{V_t^2}{0.0563 + [0.0563^2 + (0.148 + 0.22)^2]^{1/2}}$$

$$= 0.02229 \times V_t^2 \text{ at } 50 \text{ Hz}$$

(i) $V_t = 0.9711 \times 240 = 233 \text{ V}$

$$T_{max} = 0.02229 \times 233^2 = 1210 \text{ N m}$$

(ii) $V_t = 0.9711 \times 120 = 116.5 \text{ V}$

$$T_{max} = 0.02229 \times 116.5^2 = 302.7 \text{ N m}$$

(iii) $V_t = 0.9711 \times 60 = 58.3 \text{ V}$

$$T_{max} = 0.02229 \times 58.3^2 = 75.7 \text{ N m}$$

(iv) $|V_t| = \left| \frac{\frac{2}{5} \times 5.1j \times 96}{0.06 + \frac{2}{5}(0.152 + 5.1)j} \right| = 93.17 \text{ V}$

$$Z_t = \frac{\frac{2}{5} \times 5.1j(0.06 + \frac{2}{5} \times 0.152j)}{0.06 + \frac{2}{5}(0.152 + 5.1)j} = 0.0565 + 0.0606j \; \Omega$$

$$T_{max} = \frac{3 \times 4}{4 \times 2 \times 20 \times \pi} \times \frac{93.17^2}{0.0565 + \left[0.0565^2 + (0.0606 + \frac{2}{5} \times 0.22)^2\right]^{1/2}}$$

$$= 962 \text{ N m}$$

(v) $V_t = \left| \frac{\frac{1}{10} \times 5.1j \times 24}{0.06 + \frac{1}{10}(0.152 + 5.1)j} \right| = 23.16 \text{ V}$

$$Z_t = \frac{\frac{1}{10} \times 5.1j(0.06 + \frac{1}{10} \times 0.152j)}{0.06 + \frac{1}{10}(0.152 + 5.1)j} = 0.0558 + 0.0212j \; \Omega$$

$$T_{\max} = \frac{3 \times 4}{4 \times 2 \times 5 \times \pi} \times \frac{23.16^2}{0.0558 + \left[0.0558^2 + (0.0212 + \frac{1}{10} \times 0.22)^2\right]^{1/2}}$$

$$= 405.3 \, \text{N m}$$

The effect of the winding resistance in reducing the maximum torque from the ideal value at lower frequencies is seen clearly from results (i), (iv) and (v).

A variable-frequency, variable-voltage supply can be provided by using either a voltage-sourced or a current-sourced inverter as suggested by Fig. 4.29, with the voltage-sourced, pulse-width-modulated inverter being the more common. Operation at frequencies above the normal supply frequency would then be with a square or quasi-square waveform.

Where a pulse-width-modulated inverter is used, the switching frequency (f_c) used by the inverter would normally be an exact multiple of the inverter output frequency (f_i) with the ratio f_c/f_i being changed at regular speed intervals throughout the operating speed range. At low frequencies, f_c is usually held constant to avoid excess current ripple, any loss of synchronization with f_i being taken into account by the effect of the large number of pulses in each half cycle reducing the variation between cycles.

Electrical braking can be applied to an induction machine by operating it either in the generating region or plugging regions of the torque–slip characteristic. Alternatively, DC dynamic braking can be used in which the AC supply to the stator is disconnected and a controllable DC supply connected as in Fig. 4.30(a). This produces a stationary field in the airgap which results in an induced voltage, and hence induced currents, in the rotor to produce a torque which opposes the motion of the machine rotor as illustrated by Fig. 4.30(b). Additional control can be achieved by using a controlled converter as the DC supply and varying the firing angle to control the DC stator current.

For an induction machine operating from a variable frequency inverter, the output frequency of the inverter can be reduced so that the machine is operating in generating mode, returning energy to the internal DC link. If the system is provided with a regenerating converter operating in inverting mode as suggested by Fig. 4.31, then energy can be returned to the AC system.

As an alternative to the use of a regenerating converter, a braking resistor together with a controlled switch could be incorporated as in Fig. 4.32 to absorb the energy returned to the internal DC link when the motor is regenerating.

Inverter operation is described in Chapter 3.

The ratio f_c/f_i is typically varied from around 150:1 at low frequencies to 15:1 above 50 Hz.

This process is referred to as 'gear changing'.

Braking.

Fig. 4.29 Induction motor with voltage-sourced inverter.

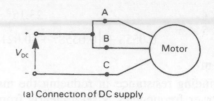

(a) Connection of DC supply

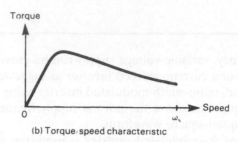

(b) Torque/speed characteristic

Fig. 4.30 DC dynamic braking of an induction motor.

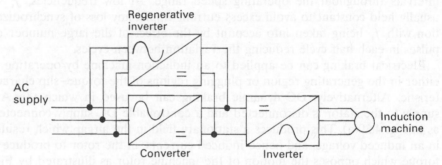

Fig. 4.31 Induction machine with a regenerative inverter.

Figure 4.33 shows a typical analogue speed controller for a voltage-sourced inverter controlled induction machine drive. The demand speed (ω_r) is compared with the speed signal provided by the tachogenerator (ω_t) to produce an error signal (ω_e). This error signal is then supplied to the regulator which provides as output (ω_{slip}) proportional to the actual speed. By limiting ω_{slip} to a value corresponding to the point of maximum torque for the machine, the motor is prevented from stalling under conditions of changing load or speed. The required synchronous speed and hence fundamental supply frequency is then determined by adding ω_{slip} and ω_t. This signal is then used to control the operation of the inverter.

A current-sourced inverter is used where rapid changes of output torque are to be avoided as the inductance in the DC link prevents rapid changes of motor current. However, the control systems for current-sourced inverter drives tend to be more complex than those for the voltage-sourced inverter because of the different operating requirements of the different types of inverter.

Figure 4.34 shows a typical controller configuration for a current-sourced inverter drive. Here, the inverter output frequency (ω_i) is compared with the

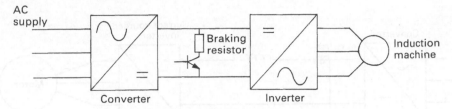

Fig. 4.32 Induction machine with braking resistor.

demand frequency (ω_r) and the resulting error signal used as input to the controller which sets the magnitude and direction of the demand torque (T_d), ensuring also that this remains within the limits for the machine. The demand torque is compared with an assessment of the actual torque (T_a) obtained from measured values of motor voltage, current and frequency to provide a second error signal which is used to adjust the operation of the inverter controller. A further error signal is also obtained by passing the demand torque to a current-limiting amplifier and comparing this with the current in the DC link. The resulting signal is then used to control the operation of the input converter.

Braking may be achieved by reducing the inverter frequency which causes the inverter input to reverse when putting the supply converter into inverting mode enables power to be returned to the AC supply.

Where operation is only required at frequencies below the frequency of the available AC supply, a cycloconverter may be used to provide a variable-frequency, variable-voltage supply. As cycloconverters are expensive in terms of the number of switching devices used and in the complexity of their control, their application tends to be limited to applications where high powers are required at the bottom end of the speed spectrum.

Cycloconverter drives.

With a wound rotor induction machine the rotor windings are connected to slip rings allowing access to the rotor circuit. This has led to the development of slip energy recovery controllers in which energy is extracted from the rotor

Slip energy recovery.

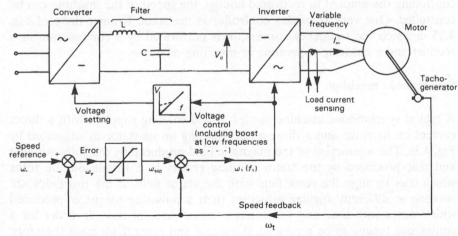

Fig. 4.33 Analogue control system for induction motor with voltage-sourced inverter.

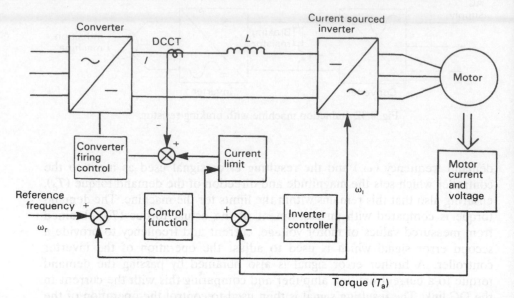

Fig. 4.34 Controller for a current-sourced inverter-fed induction motor.

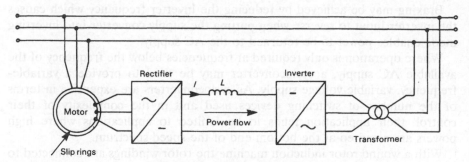

Fig. 4.35 Static Kramer system.

at slip frequency and returned to the supply via a frequency converter. By controlling the amount of recovered energy, the speed of the machine can be controlled. One version of this controller is the static Kramer drive of Fig. 4.35 in which the frequency conversion is performed by a combination of a rectifier and a converter operating in inverting mode.

Synchronous machine

A typical synchronous machine carries a field winding supplied with a direct current on its rotor and a three-phase winding on its stator as suggested by Fig. 4.36. The interaction of the magnetic field produced by the rotor winding and that produced by the stator winding results in a torque on the rotor which tries to align the rotor field with the stator field. If the two fields are rotating at different angular velocities then a pulsating torque is produced which varies with time and which has a mean value of zero. In order for a continuous torque to be produced, the stator and rotor fields must therefore be synchronized and have the same speed of rotation.

Synchronous speed $(\text{rev min}^{-1}) = n_g = 120f/p$ where f is the supply frequency and p the number of poles on the machine. Mechanical synchronous speed in mechanical radians per second $(\text{rad s}^{-1}) = \omega_{sm} = 2\pi(n_s/60) = 2\pi f \times 2/p = 2\omega/p$ where ω is the angular frequency of the supply in electrical radians per second (rad s^{-1}).

126

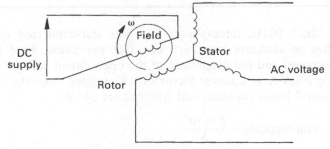

Fig. 4.36 Synchronous machine.

The speed of rotation of the rotor of the synchronous machine is therefore set by the speed of rotation of the stator field, namely synchronous speed. The synchronous speed of the synchronous machine is defined in exactly the same way as before for the synchronous speed of the three-phase induction machine.

With no applied mechanical load, then, ignoring losses, the rotor and stator fields would be in alignment. When a mechanical load is applied to the rotor shaft, the resulting torque causes the rotor to decelerate until the misalignment of the rotor and stator fields results in the production of torque required to balance the applied load.

The simple per phase equivalent circuit for a three-phase, star-connected synchronous machine is shown in Fig. 4.37 in which case the power per phase is given by

$$\text{Phase power} = P_{\text{phase}} = VI_a \cos\phi = VE_a \sin\delta/X \qquad (4.27)$$

Electrical radians = (mechanical radians) × p/2. p = number of pulses.

E_a is the open circuit voltage for the machine.

where ϕ is the phase angle between V and I_a, δ is the load angle and defines the angle between the stator and rotor fields in the air-gap of the machine

and X is the synchronous reactance of the machine.

The total power is then three times the phase power when

$$\text{Power} = P = 3VE_a \sin\delta/X = V_{\text{line}} E_{\text{a;line}} \sin\delta/X \qquad (4.28)$$

$V_{\text{line}} = \sqrt{3}\, V$ and $E_{\text{a;line}} = \sqrt{3}\, E_a$.

and

$$\text{Torque} = T = \left[\frac{p}{2\omega}\right] \frac{V_{\text{line}} E_{\text{a;line}} \sin\delta}{X} \qquad (4.29)$$

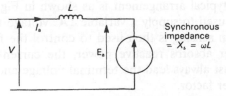

Fig. 4.37 Per phase equivalent circuit of a synchronous machine.

127

An 11 kV (line), 50 Hz, three-phase, four-pole star-connected synchronous machine has an armature reactance of 7.6 Ω per phase. Find the output mechanical power and the magnitude of the open circuit voltage E_a when it is absorbing 5 MVA at a power factor of 0.92 lagging from the AC system. The mechanical losses (windage and friction) are 54 kW.

$$VI_a = \text{Volt-amperes} = \frac{5 \times 10^6}{3}$$

Therefore

$$\text{Power} = 3 \times VI_a \cos\phi = 5 \times 10^6 \times 0.92 = 4.6 \times 10^6 \text{ W}$$

Hence

$$\text{Output mechanical power} = 4.6 \times 10^6 - 54 \times 10^3 = 4.546 \times 10^6 \text{ W}$$

From the equivalent circuit of Fig. 4.37:

$$V = E_a + I_a jX$$

where

$$I_a = \frac{5 \times 10^6}{\sqrt{3} \times 11 \times 10^3} = 262.42 \text{ A at a phase angle of } -23.07°$$

Then

$$E_a = \frac{11\,000}{\sqrt{3}} - 262.42(0.92 - 0.3919\text{j}) \times 7.6\text{j} = 5569.2 - 1834.8\text{j V}$$

in which case E_a has a magnitude of 5863.7 V.

Speed control

The speed control of a synchronous machine depends on the provision of at the machine terminals of a three-phase voltage of variable voltage and frequency such that

$$V/f = \text{constant} \tag{4.30}$$

The generation by the rotating field of a three-phase voltage in the stator winding of a synchronous machine supports the natural commutation of the inverter switching components and hence a fully-controlled, three-phase thyristor bridge operating in inverting mode can be used as the supply to the machine. A typical arrangement is as shown in Fig. 4.38 in which the supply converter is used to supply a variable DC voltage to the machine converter, rotor position sensing is then used to control the firing of the thyristors. As the converter absorbs reactive power, the current into the machine when motoring must always lead the terminal voltage and the machine operates at leading power factor.

The speed of the machine is essentially determined by the value of the DC link voltage when, in the absence of any speed feedback, there will be some

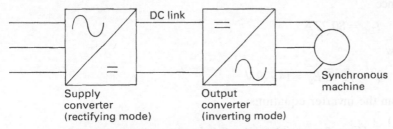

Fig. 4.38 Synchronous machine speed control.

speed regulation with increasing torque. The use of either analogue or digital controllers does enable a wide range of control strategies to be deployed.

As both converters are naturally commutated, power may flow in either direction, enabling regenerative operation to be achieved with power being returned to the supply.

On starting, natural commutation of the machine converter is not possible as no open circuit voltage is developed at zero speed. By switching the supply inverter between the rectifying and inverting modes, a series of current pulses can be supplied to the machine to produce a starting torque. As the speed of the machine increases a transition to natural commutation then takes place.

A six-pole, three-phase synchronous motor is rated at 1.2 MVA at 11 kV (line) and 50 Hz. The machine has negligible stator resistance per phase and a synchronous reactance of 10.2 Ω per phase. It is energized by a three-phase, fully controlled bridge converter operating in inverting mode from a DC link supplied by a second, similar converter. When operating with the field current set to give an open circuit voltage equivalent to 95% of the rated voltage the inverter operates with an extinction angle (ε) of 12° and draws a line current whose fundamental component has an RMS value equal to the rated current of the machine.

Worked example 4.6

Find the DC link voltage and the firing advance angle under these conditions.

$$E_a = \frac{0.95 \times 11\,000}{\sqrt{3}} = 6033.5\,V$$

RMS value of the fundamental component of line current

$$I_{a;1} = \frac{1.2 \times 10^6}{\sqrt{3} \times 11\,000} = 62.98\,A$$

when, from Chapter 2 and the current waveform for a fully controlled bridge, and using Fourier analysis,

$$\hat{I}_{a;1} =$$

$$\frac{1}{\pi}\left[\int_{-\pi}^{2\pi/3} -I_{DC}\cos\theta\,d\theta + \int_{-\pi/3}^{\pi/3} I_{DC}\cos\theta\,d\theta\right.$$

$$\left. + \int_{-\pi}^{-2\pi/3} -I_{DC}\cos\theta\,d\theta\right] = \frac{I_{DC}\,2\sqrt{3}}{\pi}$$

129

Hence

$$I_{DC} = 80.77 \text{ A}$$

Now

$$\hat{V} = \sqrt{2}\sqrt{3} E_a = 14\,779 \text{ V}$$

From the inverter equations:

$$\frac{1}{2}\hat{V}(\cos\beta + \cos\varepsilon) = \hat{V}\cos\beta + I_{DC} X_s$$

when

$$\cos\beta = 0.8667 \text{ and } \beta = 29.93°$$

The DC link voltage is then

$$V_{DC} = \frac{3}{2\pi}14\,779\,(\cos(29.93°) + \cos(12°)) = 13\,018 \text{ V}$$

Cycloconverter drive.

Cycloconverters are generally used with synchronous machines to provide a low speed, high power drive. Such drives offer high levels of efficiency with good control over a restricted speed range, typically from near zero speed to around one-third or one-half of the synchronous speed set by the supply frequency.

Synchronous motors are generally used to provide a constant speed drive, particularly at high powers they offer greater efficiency than the equivalently rated DC machine or induction machine. Applications include boiler fans and blowers, compressors, mills, conveyors and centrifuges.

Brushless machines

In a brushless machine, the field circuit is replaced by a powerful permanent magnet, eliminating the need for field winding. Construction places the magnet on the rotor as in Fig. 4.39(a) and 4.39(b), eliminating the need for a

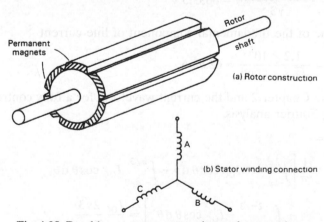

Fig. 4.39 Brushless motor construction and connection.

commutator and brushgear. The resulting machines have a number of advantages over conventional DC machines, including:

Reduced maintenance
Higher torque/volume ratio
Improved torque characteristics, particularly at high speed
Simplified protection

As a result, machines of this type have increasingly found application as servomotors in robots and machine tools.

A brushless machine can be operated either as a DC machine with the commutation performed electronically or as a synchronous machine supplied with a variable-frequency, multi-phase supply. Of the two options, the synchronous machine requires the more complex control system because of the need to produce the variable-frequency, multi-phase supply. Operation is, however, likely to be more flexible than as a DC machine.

When operated as a DC machine, the armature windings of Fig. 4.39(b) are switched in the sequence AB–AC–BC–BA–CA–CB–AB–etc. The actual point at which the switching takes place is determined by reference to rotor position, either by directly monitoring the shaft position or by using Hall effect devices to sense the magnetic field.

Brushless machines can exhibit a ripple in their output torque. The major components of this ripple are the reluctance ripple resulting from the magnetic asymmetry of the machine, a drive current ripple at a frequency of $pn/2$ and the once round ripple, at a frequency corresponding to the speed of rotation of the machine, caused by errors in the alignment of the rotor within the stator.

Electronic commutation.

p is the number of poles and n the speed in revolutions per minute.

Stepper motors

Stepper motors can be used to provide either a continuous controlled rotation or a series of discrete angular motions, making them very suitable for use in applications involving computer control of motion. A variety of different types of stepper motor are available offering a range of characteristics.

Variable reluctance stepper motor

The construction of a simple, single-stack, 12-step, variable reluctance stepper motor is shown in Fig. 4.40. In this form, the stator carries three separate windings (phases) on six equally spaced poles while the rotor has only four poles and carries no windings.

With phase AA' energized, the rotor will be pulled into the position shown in Fig. 4.40(a), with one pair of poles aligned with the phase A and A' poles on the stator. If phase AA' is now turned off and phase BB' energized, the rotor will move to the position shown in Fig. 4.40(b). A further rotation to the position of Fig. 4.40(c) occurs when phase BB' is turned off and phase CC' energized. Repeating the sequence of phases AA'–BB'–CC'–AA' etc. will cause the rotation to continue with a total of 12 steps required before the rotor returns to its original position. If this sequence in which the stator

131

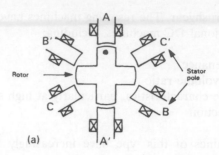

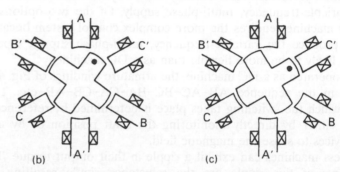

Fig. 4.40 Single-stack variable reluctance stepper motor configuration.

phases are energized is reversed to give AA′–CC′–BB′–AA′, etc. the direction of rotation of the rotor will be reversed.

A feature of this form of construction is the ability to generate intermediate steps by energizing pairs of phases together. Thus, for the motor shown, energizing phases AA′ and CC′ together can cause the rotor to assume a position between those of Fig. 4.40(a) and Fig. 4.40(b).

An alternative form of construction for a variable reluctance stepper motor is the multi-stack form of Fig. 4.41. Here, each of the stator phases is placed on a different stator stack and each stack is associated with its own rotor stack as shown. Each of these rotor stacks is aligned at some angle with respect to the other rotor stacks, enabling a smaller step angle to be achieved.

The relationship between the number of stator poles, the number of rotor poles, the number of phases and the step angle for the variable reluctance stepper motor is given by

For the machine of Fig. 4.41:
$s=3$ and $n=4 \times 2=8$.

Hence $q_s = \dfrac{360}{24} = 15°$.

$$\text{Step angle} = q_s = \frac{360}{sn} \tag{4.31}$$

where s is the number of stacks (phases)
  n is the number of teeth per stack (phase)
  = number of poles × teeth per pole

It should be noted that a variable reluctance stepper motor will only produce a torque when the stator windings are energized and that the rotor will otherwise be free to rotate if a torque is applied to the shaft.

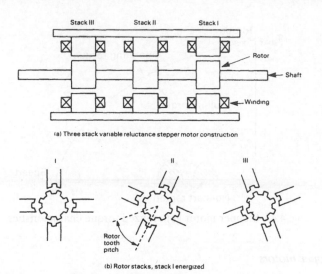

Fig. 4.41 Four-pole three-stack variable reluctance stepper motor.

Permanent magnet stepper motors

The permanent magnet stepper motor of Fig. 4.42 has, as its name suggests, a permanent magnet embedded in its rotor which increases the flux in the machine and also results in the provision of a holding or detente torque on the rotor when the stator windings are de-energized.

As with the variable reluctance stepper motor, energizing the stator windings in turn results in a stepped rotation of the rotor. The step angle is determined by the rotor and stator teeth according to Equation 4.32.

$$\text{Step angle} = \theta_s = \frac{360}{mn} \qquad (4.32)$$

m is the number of phases and n the number of teeth.

The maximum number of steps per revolution that can be achieved is limited by the construction of the machine and the size of the stator and rotor teeth that can be used.

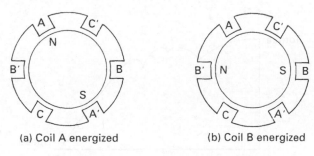

(a) Coil A energized   (b) Coil B energized

Fig. 4.42 Permanent magnet stepper motor.

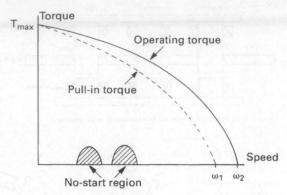

Fig. 4.43 Stepper motor generalized torque characteristics.

Hybrid stepper motors

The hybrid stepper motor combines features of both the variable reluctance and the permanent magnet stepper motor in that it combines a permanent magnet core and a toothed rotor. Hybrid stepper motors generally operate with smaller step angles than the variable reluctance and permanent magnet stepper motors and they also offer a higher torque-to-volume ratio than these. A hybrid stepper motor will also provide a detente torque when its stator windings are de-energized.

Stepper motor operation

The generalized torque–stepping-rate characteristic for a stepper motor is shown in Fig. 4.43 and exhibits a number of different regions and boundaries.

Pull-in curve.

The pull-in curve defines those combinations of torque and stepping-rate against which a motor can start or stop without losing steps. The shape of the pull-in curve is influenced by the inertia of the load that is to be started and increasing the load inertia will reduce the stepping-rate to be used if steps are not to be missed during acceleration.

Pull-out curve.

The pull-out curve defines the torque–stepping-rate relationship for the motor at steady speed and sets the limiting conditions for operation if steps are not to be lost. The shape of the pull-out curve is influenced by the nature and type of drive circuit used by the motor. It is also possible that the shape

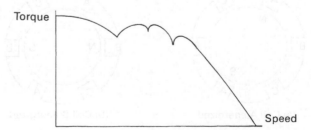

Fig. 4.44 Distorted stepper motor pull-out torque characteristic.

of the pull-out curve will be distorted in the manner of Fig. 4.44 as a result of internal resonance conditions in the motor arising from its construction.

No-start regions are, as their name suggests, regions in the torque–stepping-rate characteristic defining conditions against which the motor is unable to start. These regions are set by the construction of the motor and the type of drive used.

No-start regions.

The operation of a stepper motor is significantly influenced by the inertia of the driven load. In particular, on starting the load inertia influences the shape of the pull-in characteristic while on stopping, the rate of deceleration must be controlled in relation to the load inertia in order to prevent any overshoots occurring.

Inertia effects.

Stepper motor drive circuits

A typical drive circuit for one phase of a stepper motor is shown in Fig. 4.45. The base signals of the transistor are derived from the control circuit, which may be a dedicated integrated circuit or a microprocessor, via a suitable interface and are used to switch the transistor into the ON state.

As the performance of the stepper motor is improved by having the stator current reach its full value as rapidly as possible, the time constant (τ) of the stator circuit is reduced and the rate of rise of current increased by introducing the forcing resistor (R_1) in series with the stator to reduce the stator time constant. In order to ensure that the correct final value of stator current is obtained an increase in the applied voltage is required such that:

Time constant $\tau = \dfrac{L}{R}$.

$$I_{\text{phase}} = \frac{V_s}{R_{\text{phase}} + R_1} \tag{4.33}$$

On turn-off of the transistor, the energy stored in the magnetic field of the stator winding is dissipated through the forcing resistor and the freewheeling resistor (R_2) via the diode, causing the rapid collapse of the stator current.

Where a bidirectional current drive is required the bridge circuit of Fig. 4.46 can be used with the transistors switched in pairs (T_1–T_2–T_3–T_4).

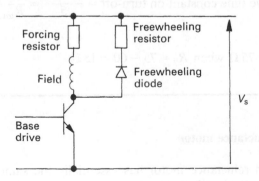

Fig. 4.45 Drive circuit for one phase of a variable reluctance stepper motor.

135

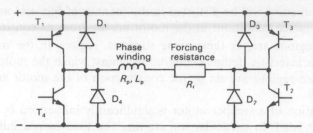

Fig. 4.46 Bidirectional current drive circuit. (Compare with Fig. 3.13, voltage-sourced single-phase inverter.)

Worked example 4.7

A variable reluctance stepper motor has the following per-phase parameters:

Inductance	=	12 mH
Resistance	=	32 Ω
Rated current	=	1.2 A

If 99% of full rated current is to be achieved within 1 ms of turn-on find the value of the forcing resistance and source voltage required. On turn-off, the current is required to collapse within 0.8 ms, hence find the value of the freewheeling resistor required.

Assuming an exponential raise of current then 99% of full rated current will be reached after a period of 5 × (system time constant).

Hence effective time constant on turn-on

$$= \frac{0.001}{5} = \frac{L}{R_{tot,1}}$$

When

$$R_{tot,1} = 60 \ \Omega$$

then

$$R_1 = 60 - 32 = 28 \ \Omega$$

The required voltage is then

$$V_s = 1.2 \times 60 = 72 \ \text{V}$$

Effective time constant on turn-off $= \dfrac{0.0008}{5} = \dfrac{L}{R_{tot,2}}$

Hence

$$R_{tot,2} = 75 \ \Omega \ \text{when} \ R_2 = 75 - 60 = 15 \ \Omega$$

Switched reluctance motor

The switched reluctance motor has essentially the same configuration as a single stack variable reluctance stepper motor but is designed for continuous rotation at high powers and torques. Referring to Fig. 4.47, continuous

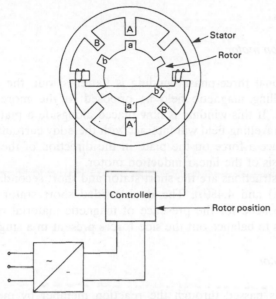

Fig. 4.47 Switched reluctance motor.

operation requires the application of current to each of the stator phases in turn at a rate which is dependent on and determined by the variation of rotor position with time. The timing of the firing of the controlling switching devices is determined by reference to the rotor position by means of either Hall effect or optical sensors. These provide input data for the control algorithms which then compute the required firing angles.

Switched reluctance motors are highly efficient machines capable of full four-quadrant operation. They are of simple construction with no rotor windings and have a flexible operating characteristic. Indeed, by varying the phase relationship of the applied current with respect to the rotor position, the characteristic of the switched reluctance motor can be shaped to a significant degree, enabling it, for instance, to assume the shaft characteristics of either a DC series or a DC shunt motor, resulting in their being used for traction applications. They do, however, require a relatively complex controller.

Speed control of switched reluctance motors

The control strategy adopted for a switched reluctance motor depends upon the operating regime. Thus for starting and low speeds, control would be applied to limit the stator current. As the machine accelerates, the stator current will fall as a back EMF appears in the windings and firing angles are adjusted to maintain the current between pre-set limits during run-up.

In the mid-range the controller provides for control of both thyristor turn-on and turn-off before finally establishing operation with controlled turn-on only at high speeds.

137

Linear motors

Linear induction motor

By fixing the plate, which then becomes the reaction member, the effect is to transfer the motion to the stator. This is the normal mode of operation.

The source of the field in a linear motor is referred to as the stator even though it might form the moving element when operated with a fixed reaction rail or secondary.

If a conventional three-phase winding is flattened out, the result will be a linearly travelling magnetic field as opposed to the more usual rotating magnetic field. If this winding is now placed alongside a plate of conducting material the travelling field will interact with the eddy currents induced in the plate to produce a force on the plate in the direction of the travelling field. This is the basis of the linear induction motor.

Typical constructions are the short stator and short secondary forms shown in Figs 4.48(a) and 4.48(b). The double-sided short stator of Fig. 4.48(b) eliminates the need for the presence of magnetic material in the secondary and also tends to balance out the side forces present in a single-sided system.

Linear DC motor

If a current is passed through the reaction member by means of brushes mounted on the rail then this current can be made to react with a DC field produced by the stator to produce linear motion.

Traction drives

Traction drives traditionally used the DC series motor with its high torque capability at low speeds. Early systems used on-board rectifiers and tap-changing transformers operating with up to 30% ripple in the motor supply current, smoothed by the inductance of the machine armature circuit. The advent of power electronic switching devices has resulted in the introduction

Additional inductance would be added in series with the armature if required.

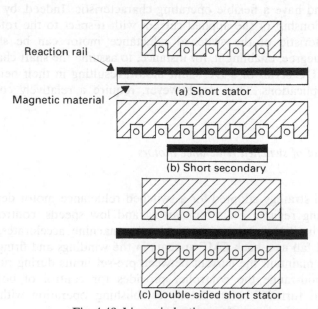

Fig. 4.48 Linear induction motors.

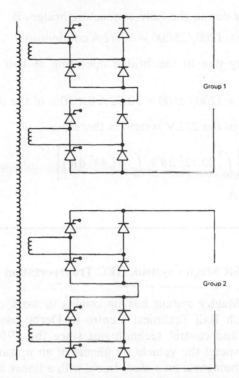

Fig. 4.49 Converter arrangement for a locomotive traction drive with separate control of the motors on each bogie.

of traction systems based separately or shunt-excited DC motors or AC motors and has increased the flexibility of operation by allowing the designer to optimize performance over the full operating envelope of the machine. The incorporation of microprocessor-based controllers has also enabled the optimization of tractive effort by allowing operation close to the limit of adhesion and has enabled greater use to be made of regenerative braking.

Figure 4.49 shows a simplified locomotive traction system using multiple bridge converters and operated from an AC supply. Reversal is achieved by means of reverse-parallel converters supplying the field windings.

The normal AC supply voltage for a traction system is single-phase at 25 kV.

Worked example 4.8

For a traction system of the type shown in Fig. 4.49, the rectifiers are each supplied with 320 V at 50 Hz from the 25 kV, 50 Hz supply. If the motor voltage in each group is three-quarters of its maximum value and the total motor current in each group is constant at 1200 A, estimate the RMS value of the current drawn from the 25 kV supply. Neglect all losses.

For three-quarters maximum voltage at the motors one bridge is fully conducting ($\alpha = 0°$) and the other is providing half its maximum output voltage ($\alpha = 90°$).

Transformer ratio = 25 000/320 = 78.125

139

Current in primary due to the fully conducting bridges is

$$I_{p1} = 2 \times 320 \times 1200/25\,000 = 30.72\,\text{A continuous}$$

Current in primary due to the bridge operating at half maximum output voltage is

$$I_{p2} = 2 \times 320 \times 1200/25\,00 = 30.72\,\text{A for 50\% of the time.}$$

The RMS current in the 25 kV system is therefore

$$I_{25\text{kV}} = \left[\frac{1}{2\pi} \left(\int_0^\pi 30.72^2 \, \mathrm{d}\theta + \int_0^\pi 61.44^2 \, \mathrm{d}\theta \right) \right]^{1/2}$$
$$= 48.57\,\text{A}$$

Case study – the BR Maglev system, GEC Transportation Projects

The British Rail Maglev system has its origins in work carried out in the 1970s at the British Rail Technical Centre in Derby based upon developments in magnet and control technologies since the 1950s. The approach adopted was to suspend the vehicle by means of an upward attractive force produced by electromagnets on a steel track, with a linear induction motor to provide the propulsive force. Though this means of suspension is inherently unstable it proved possible to control the system electronically to maintain a mean gap of 15 mm.

On the basis of this development work the BR Maglev system was chosen for the passenger link between Birmingham International Airport and the National Exhibition Centre (NEC) in competition with other, more conventional, forms of transport. The system, details of which are given in Table 4.2, is intended to provide a shuttle service with a design capacity of 2215 passengers per hour over a track length of 623m. Two-car trains each with a maximum payload of three tonnes run on parallel tracks with a maximum speed of 42 km h^{-1}. The whole of the system can be operated from a single control desk. Provision is made for features such as unmanned operation, programmed service frequency or passenger call service. Safety systems are designed on a fail-to-safe basis with appropriate back-up systems.

The consortium operated under the name of 'People Mover Group'.

The project was carried out by a consortium of British companies with GEC Transportation Projects as the project managers. Construction began in 1981 and the system entered partial public service on 7 August 1984 and full service soon thereafter.

The track

The track system adopted is shown in Fig. 4.50 and is carried on a concrete structure 5 m above ground. Alignment incorporates two curves of about 50 m radius and a gradient of 1.5%. A removable track section is included for vehicle removal. The track itself is constructed of steel sleepers carrying the laminated steel support rails, a pair of aluminium power supply rails at 600 V and the reaction rail for the linear induction motor. Also included in the

Table 4.2 Birmingham International Airport / NEC Maglev System Details

System Design		System Performance	
Configuration	Dual independent tracks on an elevated guideway, Stations at each end, substation in middle	Acceleration	$1.24\,\mathrm{m\,s^{-2}}$ max.
		Emergency braking rate	$2\,\mathrm{m\,s^{-2}}$
		Clamp-up deceleration	$5\,\mathrm{m\,s^{-2}}$
		Jerk	$1.25\,\mathrm{m\,s^{-3}}$ max.
Track length	623 m buffer-to-buffer	Speed	$15\,\mathrm{m\,s^{-1}} = 54\,\mathrm{km\,h^{-1}}$ $= 34\,\mathrm{mile\,h^{-1}}$
Passenger flow	190 per direction per 15 min	Vertical ride	0.045g
		Lateral ride	0.033g (0.08g on curve)
Track capacity	397 passengers per 15 min or 274 passengers + luggage per direction per 15 min	Longitudinal ride	0.033g
		Lateral natural frequency	1.5 Hz
Car capacity	34 standing, 6 seated	Stopping accuracy	$\pm 100\,\mathrm{mm}$
		Max roll angle	$\pm 2.6°$
Frequency	8 train trips per 15 min	Max pitch angle	$\pm 1.26°$
		Max yaw angle	$\pm 1.3°$
Journey time	100 s	Emergency braking distance	50 m max. (ice-free)
Dwell time	44 s nominal Adjustable 5 to 90 s	Energy consumption	2 k W h per vehicle journey
Floor space	98 $\mathrm{m^2}$ per direction per 15 min		
Noise level	60 to 66 dBA 3 m from track at $13\,\mathrm{m\,s^{-1}}$, 75 dBA inside vehicle		

Track requirements

Design life	50 years	Max gradient	1.5%
Banking	None	Curve, horizontal, min.	40 m radius
Level	$\pm 5\,\mathrm{mm}$, $-10\,\mathrm{mm}$ abs.		
Rate of change	1 in 500 on 10 mm chord	Curve, vertical, min.	1400 m radius
		Alignment	$\pm 4\,\mathrm{mm}$, $-4\,\mathrm{mm}$, 10 mm chord

Vehicle

Length	6.00 m external; 5.54 m internal
Width	2.25 m external; 1.85 m internal
Height	3.50 m external; 2.23 m internal
Weights	5000 kg empty; 8000 kg loaded (3000 kg payload)

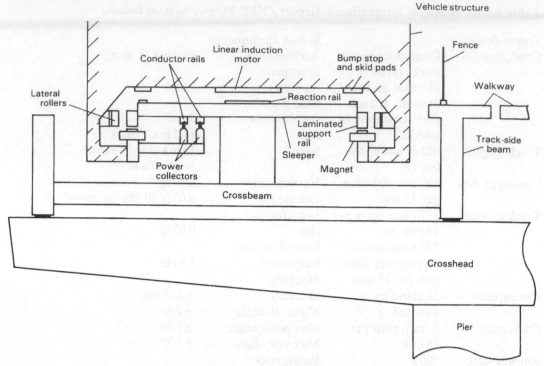

Fig. 4.50 Maglev track system.

track system are cable loops for communication between the vehicle and trackside and marker plates for overspeed detection. Docking at the stations is to within ± 100 mm with a maximum gap between the platform and vehicle of 55 mm.

Power supply

The track power supply of 600 V, 700 A, DC is fed from the incoming 11 kV AC system by transformer/rectifiers, with separate supplies for each track. To prevent stray currents and associated interference and corrosion problems, single-point earthing at the substation is used. The vehicle collects power via metallized carbon collector shoes running in channels on the conductor rails.

On board the vehicles, auxiliaries are supplied from a 3 kW, 600/48 V DC/DC converter with a 24 A h lead–acid battery backup. Load shedding is used on a failure of the main supply, allowing sensitive loads to be maintained for about 20 min.

Control and communication

Control and communication is based upon the following systems:

(a) The Automatic Train Protection (ATP) system dealing with safety functions.
(b) The Automatic Train Operation (ATO) system performing the driving functions.

142

(c) The supervision system interfacing with the control-room operator.

(d) The communications system providing a link between the operator and the passengers.

Operations can either be continuous with running on one or both tracks with a preset dwell time at each station or on-demand when the vehicles run on response to passenger requests. Starting is under control of the central or local scheduling logic with running operation controlled by the ATO computer.

Suspension and guidance

A magnetic suspension system must control the inherently unstable suspension gap in relation to ride quality, taking into account factors such as the interaction between the suspension control and the resonances of both the supporting structure and the vehicle.

The suspension system uses eight magnets mounted in pairs at the vehicle corners, the magnets of each pair being laterally offset on opposite sides of the rail centre line to provide guidance. The magnets operate with a nominal gap of 15 mm and an airgap flux density of 0.8 T. The average power for levitation and control is about $3\,\mathrm{kW\,t^{-1}}$ giving a magnet lift/weight ratio of the order of 11.7:1 with a 15 mm gap.

The magnet pairs are driven by the DC chopper of Fig. 4.51, together for vertical control and guidance and differentially for lateral damping. Each of the two-quadrant bridge circuits is switching in anti-phase at 1 kHz. Chopper efficiency is 97%. On turn-on the base current is held at about 2 A for about 6 μs before falling to a lower value. There is also a high initial collector current pulse due to the discharge of the snubber capacitor. During turn-off an *LC* circuit is used to provide a high peak reverse current in the base circuit to optimize turn-off performance. A complete turn-on or turn-off takes about 8 μs.

Protection of the power transistors is included within the base drive circuit while separate crowbar thyristors are used to protect both the power supply and the levitation magnets.

Stability of the suspension system is obtained by ensuring that the magnet currents can be adjusted more rapidly than changes of magnetic force

Conditions for rapid turn-on and turn-off of transistors are given in Chapter 1.

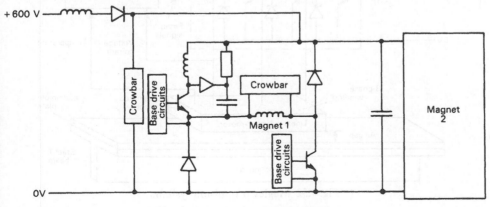

Fig. 4.51 Levitation power system.

produced by the magnets can occur. Vertical damping is obtained by using a signal proportional to the rate of change of the airgap. The restoring forces introduced by offsetting the magnet pairs mean that the guidance system is inherently stable but requires some lateral damping. This is achieved by differentially exciting each pair of magnets to ensure the resulting vertical force is constant. The additional lateral force ensures stability.

Propulsion and braking

The linear induction motor system was supplied by Brush Electrical Machines Ltd.

The propulsion system is shown in Fig. 4.52 and uses an inverter-driven, short-stator, single-sided, axial-flux linear induction motor mounted centrally under the vehicle. The reaction rail is a steel beam capped by an aluminium plate and fixed to the suspension rail. The motor develops a tractive thrust of $4\,kN$ from rest to $15\,m\,s^{-1}$ with a maximum continuous rating of $2\,kN$ at $15\,m\,s^{-1}$ at an operational air gap for the motor of $20\,mm$.

The inverter is of the pulse-width-modulated type using transistor switching modules and is rated at $240\,kVA$ continuously or $325\,kVA$ ($450\,A$) for $60\,s$. The output frequency range is 0 to $45\,Hz$ at power factors in the range of 0.5 to 0.9, lagging or leading.

Waveform generation, inverter control and protection are achieved by means of a microprocessor-based controller. The inverter control will respond to a demand by the ATO computer, taking into account vehicle speed, line voltage, motor current, motor airgap and jerk limit. The slip is adjusted to give either a motoring or a braking force. The velocity profile is matched to the track alignment and the required acceleration and deceleration rates.

Cooling of the motor is achieved by air ducted over the core and winding with a blower automatically cutting-in if the winding exceeds a specified

Jerk is the rate of change of acceleration (d^2 (velocity)/dt^2 $=m\,s^{-3}$). The jerk limit applied for the Maglev system is $0.6\,m\,s^{-3}$.

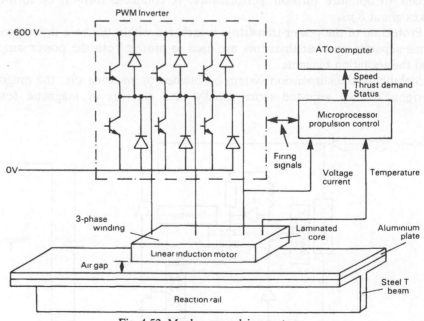

Fig. 4.52 Maglev propulsion system.

temperature. The inverter uses natural air cooling via fins on the exterior of the vehicle.

(Based on: Nenadovic, V. and Riches, E.E. (1985). Maglev at Birmingham Airport: from system concept to successful operation. *GEC Review* **1** (1). Reproduced with permission of GEC.)

Problems

4.1 Referring to the section in Chapter 2, 'Converters with voltage bias', the DC motor is operating from a 240 V (RMS) supply and is developing a mean torque of 2 N m. The thyristors are being fired at $\alpha = 100°$ and the armature current continues for 45° beyond the voltage zero. If the motor has a torque constant ($K\phi$) of $1\,\mathrm{N\,m\,A^{-1}}$ and an armature resistance of 4.8 Ω, what will be the motor speed under these conditions? Neglect all mechanical losses in the motor.

4.2 A separately excited DC motor is supplied from a 415 V, 50 Hz supply by a fully-controlled, single phase bridge converter. The motor parameters are: armature resistance 5.4 Ω, armature inductance 48 mH, torque constant $1.4\,\mathrm{N\,m\,A^{-1}}$ and voltage constant ($K\phi$) 1.4 V per $\mathrm{rad\,s^{-1}}$. The motor is operated with closed-loop control and is operating at a speed of 1800 rev min^{-1} with a firing angle of 50°. What will be the net output torque? Windage and friction are constant at an equivalent torque of 0.6 N m.

4.3 If the machine and converter of Problem 4.2 are configured to provide regeneration, what will be the mechanical input torque when the machine is being driven at 2000 rev min^{-1} and the converter is operating with a firing advance angle of 36°? The windage and friction losses are the same as in Problem 4.2.

4.4 A DC motor is operating at constant speed from a single-phase bridge converter. With reference to the section in Chapter 2, 'Converters with voltage bias' and Fig. 2.35, sketch the armature current and voltage waveforms for the following conditions:
(a) Cut-off angle (ψ) > firing angle (α) > ($\sigma - \pi$)
(b) $\psi > \alpha$ and ($\sigma - \pi$) > α
(c) $\psi > \alpha$ and ($\sigma - \pi$) = ψ

4.5 A DC shunt-wound machine has the parameters: armature resistance 0.78 Ω, armature inductance 16 mH, torque constant $2.1\,\mathrm{N\,m\,A^{-1}}$ and voltage constant 0.209 V per rev min^{-1}. It is supplied by a fully controlled, three-phase bridge converter from a 220 V (line), 50 Hz AC source. Find expressions for the armature current and hence determine the values for mean torque, speed, etc., as appropriate for the following conditions:

(a) Motoring with a firing angle of 22.5°, a continuous current and 70 N m torque.
(b) Generating with a firing advance angle of 22.5°, a continuous current and 70 N m torque.

145

(c) Motoring with a firing angle of 67.5° at a speed of 720 rev min⁻¹.

(d) Generating at a firing advance angle of 67.5° at a speed of 720 rev min⁻¹.

4.6 A 415 V (line), 50 Hz, four-pole, three-phase induction motor develops an output shaft power of 5.2 kW when running at 1430 rev min⁻¹. The mechanical losses are 250 W. Calculate the rotor copper loss and the power transferred across the airgap from stator to rotor. Iron losses may be neglected.

4.7 An inverter is used to supply a 415 V (line), 50 Hz, four-pole, three-phase induction machine. Find the approximate output of the inverter when the machine is operated at 600, 900 and 1200 rev min⁻¹.

4.8 A three-phase, four-pole induction motor has the equivalent circuit shown and is supplied by an inverter at 45 Hz with a voltage of 197.4 V per phase. Estimate the speed and torque of the machine when the slip is 0.04.

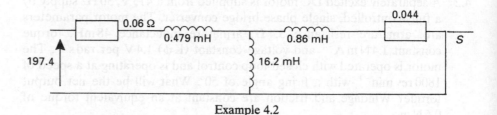

Example 4.2

5
Applications II – power supplies

□ To examine the operation of switched-mode and series-resonant power supplies.

□ To consider the deployment and use of uninterruptable power supplies.

Objectives

Switched-mode power supplies

The switched-mode power supply (SMPS) was originally developed by NASA in the 1960s as a compact DC source for space vehicles which could provide a constant, and controllable, output voltage with minimal ripple over a wide range of load current and which could operate from a variable source. From the mid-1970s onwards, the development of the SMPS was rapid and they now account for the great majority of the power supplies being produced with application ranging from electronics and computers to aerospace and automobiles.

An SMPS is essentially a DC chopper with a filtered and smoothed output. The magnitude of the output voltage is controlled by varying the mark–space ratio of the chopper using either pulse-width control at constant frequency or variable frequency at constant pulse width. Figure 5.1 shows the basic configuration of the step down, step-up and flyback converter forms of the basic SMPS.

Also referred to as the buck, boost and buck/boost converters.

Step-down converter

Referring to Fig 5.1(a) and assuming that the transistor T is initially off with zero current in the inductor ($i_L = 0$) and the output voltage (V_o) is at its nominal value then the capacitor C will discharge through the load resistor (R_o), causing the output voltage to fall below the nominal value.

The output voltage of the step-down converter is less than the supply voltage.

Turning transistor T on allows a current to flow through the inductor into the parallel combination of the capacitor and load. During this period the voltage across the inductor is expressed by

If the saturation voltage (V_{sat}) of the transistor is included then

$$L\frac{di_L}{dt} = V_s - V_o \tag{5.1}$$

$$L\frac{di_L}{dt} = V_s - V_o - V_{sat}$$

For a conduction period t_{on} this gives the solution for i_L of the form

$$i_{L,max} = \left[\frac{(V_s - V_o)}{L}\right]t_{on} \tag{5.2}$$

When transistor T is turned off, the magnetic field in the inductor starts to collapse, producing a reverse voltage the forward biases diode D, allowing the

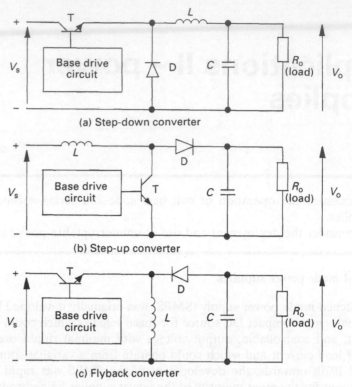

(a) Step-down converter

(b) Step-up converter

(c) Flyback converter

Fig. 5.1 Converter forms of a switched-mode power supply.

inductor current to decay. During this period the voltage across the inductor is given by

If the forward voltage drop (V_D) of the diode is included then

$$L\frac{\mathrm{d}i_L}{\mathrm{d}t} = -(V_o + V_D).$$

$$L\frac{\mathrm{d}i_L}{\mathrm{d}t} = -V_o \qquad (5.3)$$

For an off period of t_{off} and assuming an initial current of $i_{L,\,max}$, this gives

$$i_L = i_{L,\,max} - \left[\frac{V_o}{L}\right]t_{off} \qquad (5.4)$$

Discontinuous mode.

In order to satisfy the initial conditions set out earlier the current in the inductor during the off period must fall to zero, in which case

$$\left[\frac{(V_s - V_o)}{L}\right]t_{on} = \left[\frac{V_o}{L}\right]t_{off} \qquad (5.5)$$

when

Including diode and transistor on voltages

$$\frac{t_{on}}{t_{off}} = \frac{V_o + V_D}{V_s - V_o - V_{sat}}$$

$$\frac{t_{on}}{t_{off}} = \frac{V_o}{V_s - V_o} \qquad (5.6)$$

The mean output voltage is then

$$V_o = V_s\left[\frac{t_{on}}{t_{on} + t_{off}}\right] \qquad (5.7)$$

If the output voltage is to remain essentially constant, the average current through the inductor must be equal to the output current for a complete cycle since the voltage–time product at the inductor must be the same for both the on and off periods. Hence

$$I_o = \frac{i_{L,\,max}}{2} \qquad (5.8)$$

The peak current through the inductor is also the peak current through the transistor.

The value of t_{off} is determined by the time required for the current in the inductor to fall to zero, in which case the maximum on period, $t_{on,\,max}$, can be found by reference to the minimum switching frequency f_{min}, when

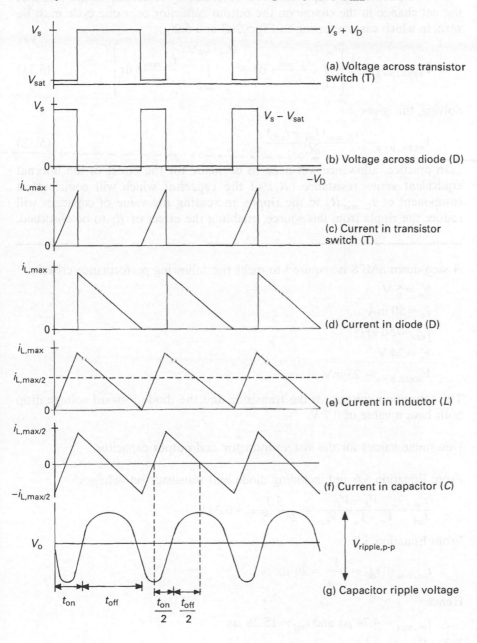

(a) Voltage across transistor switch (T)

(b) Voltage across diode (D)

(c) Current in transistor switch (T)

(d) Current in diode (D)

(e) Current in inductor (L)

(f) Current in capacitor (C)

(g) Capacitor ripple voltage

Fig. 5.2 Step-down converter waveforms.

$$f_{min} = \frac{1}{t_{on,max} + t_{off}} \tag{5.9}$$

Incorporating the saturation voltage for the transistor gives

The minimum value of inductance required can now be found by reference to the maximum on period and the peak switch current allowed when

$$L_{min} = \left[\frac{V_s - V_o - V_{sat}}{i_{L,max}}\right] t_{n,max}$$

$$L_{min} = \left[\frac{V_s - V_o}{i_{L,max}}\right] t_{on,max} \tag{5.10}$$

The output voltage ripple can now be found by reference to the fact that the net change in the charge on the output capacitor over one cycle must be zero, in which case, referring to Fig 5.2(f) and 5.2(g):

$$V_{ripple,p=p} = \frac{1}{C}\left[\int_0^{t_{on}/2} \frac{i_{L,max}}{t_{on}} t \, dt + \int_{t_{on}/2}^{(t_{on}+t_{off})/2} \frac{i_{L,max}}{t_{off}} t \, dt\right] \tag{5.11}$$

Solving, this gives

$$V_{ripple,p=p} = \frac{i_{L,max}(t_{on} + t_{off})}{8C} \tag{5.12}$$

In practice, allowance will need to be made for the effect of the internal equivalent series resistance (R_c) of the capacitor which will contribute a component of $i_{L,max} R_c$ to the ripple. Increasing the value of capacitor will reduce the ripple from this source, enabling the effect of R_c to be absorbed.

Worked example 5.1

A step-down SMPS is required to meet the following performance criteria:

$V_o = 5$ V

$I_o = 50$ mA

$f_{min} = 50$ kHz

$V_s = 24$ V

$V_{ripple,p=p} = 25$ mV

The saturation voltage for the transistor and the diode forward voltage drop both have a value of 0.7 V.

Determine values for the system inductor and output capacitor.

From Equation 5.6 and including diode and transistor on-voltages:

$$\frac{t_{on}}{t_{off}} = \frac{V_o + V_D}{V_s - V_o - V_{sat}} = \frac{5.7}{24 - 5.7} = 0.311$$

From Equation 5.9:

$$t_{on,max} + t_{off} = \frac{1}{f_{min}} = 20 \ \mu s$$

Hence

$$t_{on,max} = 4.74 \ \mu s \text{ and } t_{off} = 15.26 \ \mu s$$

From Equation 5.8:

$$i_{L,max} = 100 \text{ mA}$$

The required value of inductance can now be found using Equation 5.10:

$$L_{min} = \left[\frac{V_s - V_o - V_{sat}}{i_{L, max}} \right] t_{on, max} = \left[\frac{18.3}{100 \times 10^{-3}} \right] \times 4.74 \times 10^{-6} = 867.4 \ \mu H$$

The output capacitor can be found using Equation 5.12:

$$C = \frac{i_{L, max}(t_{on} + t_{off})}{8 V_{ripple, p=p}} = \frac{100 \times 10^{-3} \times 20 \times 10^{-6}}{8 \times 25 \times 10^{-3}} = 10 \ \mu F$$

A capacitor of this value would typically have an equivalent series resistance of around 0.2Ω to 0.3Ω which would result in a peak-to-peak ripple voltage of between 20 mV and 30 mV. A larger capacitor would therefore be required in practice.

Step-up converter

The step-up converter is shown in Fig. 5.1(b) and is intended to provide an output voltage (V_o) greater than the supply voltage V_s such that

$$V_o = V_s \left[\frac{t_{on}}{t_{off}} + 1 \right] \tag{5.13}$$

Assuming that the current through the inductor is initially at zero with the transistor T turned off then the load current will be supplied by the output capacitor C. At the point at which the output voltage reaches its permitted lower level, transistor T is turned on, allowing current to flow through the inductor L. During the on period (t_{on}):

$$L \cdot \frac{di_L}{dt} = V_s \tag{5.14}$$

when

$$i_L = \left[\frac{V_s}{L} \right] t \tag{5.15}$$

Once the required current has been reached in the inductor, transistor T is turned off and the magnetic field of the inductor will start to collapse, forward biasing diode D and driving a current into the output capacitor and load as shown in Fig. 5.3(f). During this period the current in the inductor is expressed by

$$L \frac{di_L}{dt} = V_s - V_o \tag{5.16}$$

Assuming a current of $I_{L, max}$ in the inductor at the time of switching then

$$i_L = i_{L, max} = \frac{V_o - V_s}{L} t \tag{5.17}$$

In order to satisfy the initial conditions on transistor turn-on defined above, the current in the inductor must fall to zero during the off period. Assuming an on period of duration t_{on} and an off period of duration t_{off}, this gives, from Equations 5.15 and 5.17:

If the saturation voltage, V_{sat}, of the transistor is included then

$$L \frac{di_L}{dt} = V_s - V_{sat}$$

If the diode forward voltage drop V_d is included the relationship becomes

$$L \frac{di_L}{dt} = V_s - V_o - V_d$$

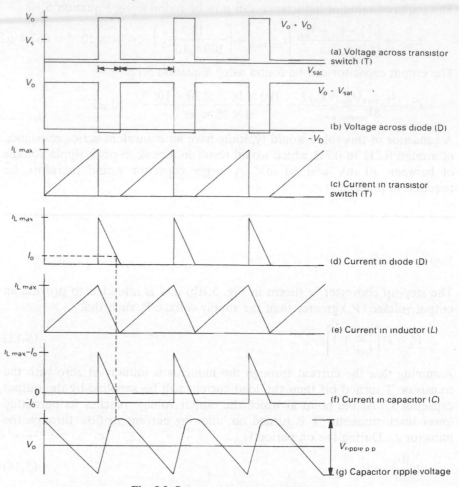

Fig. 5.3 Step-up converter waveforms.

The waveform labels shown in the figure:

$V_o + V_D$ — (a) Voltage across transistor switch (T)

$V_o - V_{sat}$ — (b) Voltage across diode (D)

— (c) Current in transistor switch (T)

— (d) Current in diode (D)

— (e) Current in inductor (L)

— (f) Current in capacitor (C)

$V_{ripple\,p\,p}$ — (g) Capacitor ripple voltage

If the transistor and diode voltage drops are included this gives

$$\frac{t_{on}}{t_{off}} = \frac{V_o + V_D - V_s}{V_s - V_{sat}}$$

$$\frac{t_{on}}{t_{off}} = \frac{V_o - V_s}{V_s} \tag{5.18}$$

During one cycle the net change in charge on the output capacitor must be zero. Hence, referring to Fig. 5.3(f):

$$t_1 = \left[\frac{i_{L,\,max} - I_o}{i_{L,\,max}}\right] t_{off} \tag{5.19}$$

and

$$Q_+ = Q_- \tag{5.20}$$

when

$$\left[\frac{i_{L,\,max} - I_o}{2}\right] t_1 = I_o t_{on} + \frac{1}{2} I_o (t_{off} - t_1) \tag{5.21}$$

Substituting for t_1 from Equation 5.19 and solving gives

$$i_{L,\,max} = 2 I_o \left(1 + \frac{t_{on}}{t_{off}}\right) \tag{5.22}$$

The value of inductance required can now be found from the peak current and the maximum permitted on-time, $t_{on, max}$:

$$L_{min} = \left[\frac{V_s}{i_{L, max}} \right] t_{on, max} \qquad (5.23)$$

The output capacitor value can be obtained as before by reference to the allowable peak-to-peak ripple voltage ($V_{ripple, p-p}$). Referring to Fig. 5.3(f) and 5.3(g):

$$V_{ripple, p-p} = \frac{1}{C} \int_0^{t_1} \left[\frac{i_{L, max} - I_o}{t_1} \right] t \; dt \qquad (5.24)$$

Solving and substituting for t_1 as before:

$$V_{ripple, p-p} = \frac{(i_{L, max} - I_o)^2 t_{off}}{2 i_{L, max} C} \qquad (5.25)$$

Flyback converter

The flyback form of converter is shown in Fig. 5.1(c) and provides an inverted voltage output at a magnitude which can be either greater or less than the supply voltage V_s. The operation of the flyback SMPS is essentially similar in detail to that of the step-up SMPS with

$$|V|_o = V_s \left[\frac{t_{on}}{t_{off}} \right] \qquad (5.26)$$

Operation is illustrated by Fig. 5.4

Cuk converter

The Cuk converter of Fig. 5.5 is a variation on the flyback converter and is similarly capable of providing an output voltage which is either greater than or less than the supply voltage, also at reverse polarity. In operation, with transistor T off, current flows from the source via inductor L_1 and diode D to charge capacitor C_1 while the load is supplied from the output capacitor C_2. When transistor T is turned on, C_1 discharges via the transistor and inductor L_2, transferring energy to the output capacitor C_2. Assuming a linear variation of all currents and voltages in the circuit and ignoring the ripple in the output voltage then the charge supplied to capacitor C_1 from the source with transistor T off must equal that supplied to capacitor C_2 when the transistor is on. In which case

$$I_{c1} t_{off} = I_{c2} t_{on} \qquad (5.27)$$

Also, for an ideal system:

$$\text{Power from source} = V_s I_{c1} = V_o I_{c2} = \text{Power to load} \qquad (5.28)$$

when

$$|V_o| = V_s \left(\frac{t_{on}}{t_{off}} \right) \qquad (5.29)$$

Compare Equation 5.29 with Equation 5.26 for the flyback converter.

153

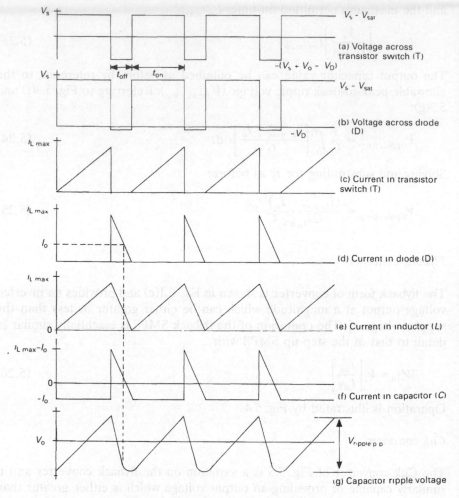

(a) Voltage across transistor switch (T)

$V_s - V_{sat}$

$-(V_s + V_o - V_D)$

t_{off} t_{on}

(b) Voltage across diode (D)

$V_s - V_{sat}$

$-V_D$

(c) Current in transistor switch (T)

$I_{L\,max}$

(d) Current in diode (D)

$I_{L\,max}$

I_o

(e) Current in inductor (L)

$I_{L\,max}$

0

(f) Current in capacitor (C)

$I_{L\,max} - I_o$

0

$-I_o$

(g) Capacitor ripple voltage

V_o

$V_{ripple\,p\,p}$

Fig. 5.4 Flyback converter operation.

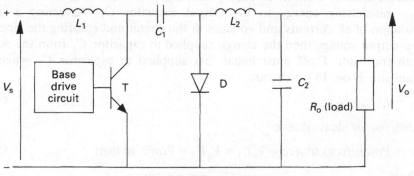

Fig. 5.5 Cuk converter.

154

Isolation

The basic circuits considered so far contain no isolation. By incorporating the necessary inductance into a transformer winding, isolated forms of the basic converter circuits are obtained as in Fig. 5.6. These are identical in operation to the forms shown in Figs. 5.1(c) and 5.5 except that the presence of the transformer provides both isolation and a range changing capability.

Other converter forms

Through using minimal numbers of components, the basic converters considered above make a somewhat inefficient use of the switching device and have relatively high levels of noise in their output. As a result, applications tend to be limited to power levels of a few hundred watts, and alternative configurations have therefore been developed for high power ratings.

Cuk, S. (1984). Survey of switched mode power supplies. *IEE Conference on Power Electronics and Variable Speed Drives*, IEE Publication 234, pp. 83-94.

The full-bridge converter of Fig. 5.7 was developed for use at power ranges of the order of a few kilowatts and uses a pulse-width-modulated (PWM) strategy to produce an output voltage of

Full-bridge converter.

$$V_o = 2nV_s \left[\frac{t_{on}}{t_{on} + t_{off}} \right] = 2nV_s \delta \qquad (5.30)$$

where n is the turns ratio of the transformer stage and δ is the duty cycle of the PWM wave form and has a maximum value of 0.5.

At lower power levels the full-bridge converter can be simplified by replacing two of the transistors by capacitors as in Fig. 5.8. The capacitors

Half-bridge converter.

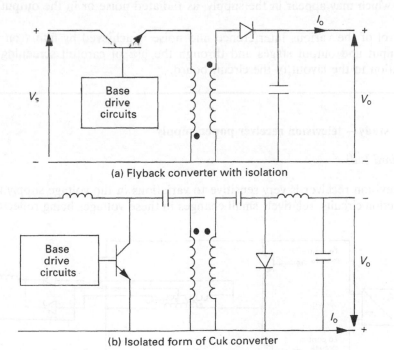

(a) Flyback converter with isolation

(b) Isolated form of Cuk converter

Fig. 5.6 Isolated forms of switched-mode power supplies.

155

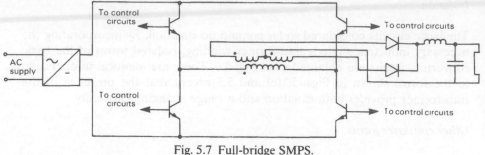

Fig. 5.7 Full-bridge SMPS.

now act as a potential divider limiting the voltage excursions to half those for the full bridge.

The use of the push–pull form of converter of Fig. 5.9 eliminates the need for capacitors and results in voltage excursions of $2V_s$ at the primary winding.

The overall size of an SMPS is a function of the frequency of operation. While the majority of SMPS operate with frequencies in the range 20 kHz to 50 kHz or 100 kHz, power supplies operating with frequencies to the megahertz range have been developed for aircraft systems using MOS switches.

If a comparison is made between a conventional, linear power supply and an SMPS of the same rating, the SMPS will be of smaller size, lighter weight and higher efficiency. It will also be less sensitive to variations in the input voltage level. On the debit side, an SMPS will tend to have a larger output ripple, a reduced dynamic response and a worse regulation. An SMPS may also be a source of both electromagnetic and radio frequency (RF) interference which may appear in the supply, as radiated noise or in the output.

Control of the various interference and noise is achieved by filters on both the input and output stages and through the use of carefull screening and attention to the layout of the circuit board.

Case study – television receiver power supply

(Mullard Ltd)

A television receiver is very sensitive to variations in the voltage supply to its deflection circuits, relatively small changes in these voltages being reflected in

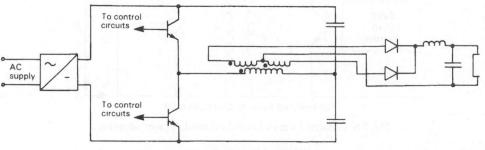

Fig. 5.8 Half-bridge SMPS.

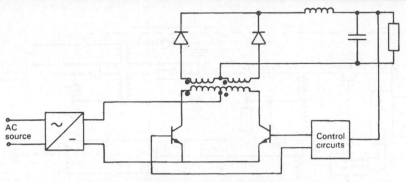

Fig. 5.9 Push–pull SMPS.

visible changes in picture size. For this reason the SMPS with its ability to provide a stable output voltage over a wide range of loads and input voltage conditions has been adopted for use in television receiver horizontal deflection circuits.

Figure 5.10 shows the block diagram of an SMPS supplying the horizontal deflection circuit, east–west raster correction circuit, line deflection coil, sound output stage and some auxiliary circuits. The SMPS is of the single transistor flyback type and has the performance requirements set out in Table 5.1.

The decision of the SMPS must be such as to ensure that the operating load line of the output switching transistor is maintained within its SOA under all operation conditions. Additionally, action must be taken to prevent any radio frequency interference (RFI) generated by the SMPS causing visible interference on the screen. Referring to the SMPS output circuit of Fig. 5.11 the RFI can be controlled by the connection of the output switching transistor to the upper end of the transformer primary and by the inclusion of capacitor C_1. The first of these measures maintains the stray capacitance at the collector of the output switching transistor at a constant potential, reducing switching spikes, while the inclusion of C_1 reduces the dV/dt values and minimizes the length of the radiation loop.

Further measures required to maintain the output switching transistor within its SOA are the inclusion of the inductance L to limit the peak current in C_1 on turn-on, the addition of the D_1–R_1 combination to damp out the LC_1 oscillations and the inclusion of capacitor C_2 to minimize the peak voltage across the transistor by passing the peak current. Finally, the D_2–R_2–C_3 network is incorporated to limit overshoot due to the transformer leakage inductance on turn-off.

The base drive to the output switching transistor provides a fast rising base current for rapid turn-on to minimize switching losses. Following turn-on the transistor is maintained in saturation for the duration of its conduction period. Losses on turn-off are minimized by ensuring that a reverse base current is maintained once the collector current starts to decrease. The base drive circuit must also be able to hold the output switching transistor off following remote turn-off or operation of protection systems. It also provides the interface between the control circuits and the output switching transistor.

Transistor turn-on and turn-off is discussed in Chapter 1.

157

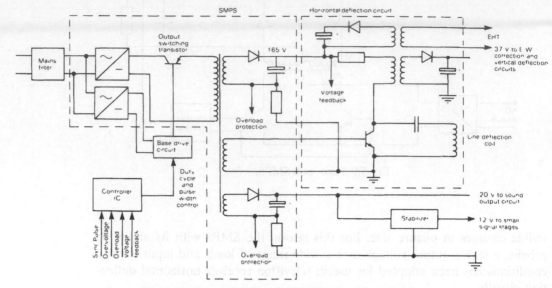

Fig. 5.10 Television receiver SMPS, simplified circuit.

Overload protection is provided in each of the output transformer secondaries by sensing the voltage across a reference resistor connected to earth. This voltage is taken to the controller integrated circuit which will then shut down the SMPS on detecting an overload condition. The collector current of the output switching transistor can be measured directly by including a current-sensing transformer in series with the transistor. The output from this transformer is converted to a voltage signal and taken to the overvoltage detection circuit of the controller integrated circuit.

The operation of the SMPS can result in the following types of interference:

(a) **Visible interference** – the result of spurious radiated signals into the tuner and/or i.f. amplifiers of the receiver. The principal source of this

Table 5.1 Requirements of Switched-mode Power Supply for Television Receiver

Mains voltage range	V_{mains}	187–265 V AC (220 V, +20%, −15%)
Main output voltage	V_0	165 V DC
Minimum load on main output	$P_{0, min}$	60 W
Maximum load on main output	$P_{0, max}$	120 W
Losses in SMPS output circuit	P_{loss}	10 W
Minimum auxiliary load	$P_{a, min}$	10 W
Maximum auxiliary load	$P_{a, max}$	20 W
Base drive auxiliary load	P_{drive}	4 W
Stabilization of main output		Better than 3%
Ripple level 50/100 Hz hum		Less than 0.2%
Switching frequency		15 625 Hz ($T = 64\ \mu s$)

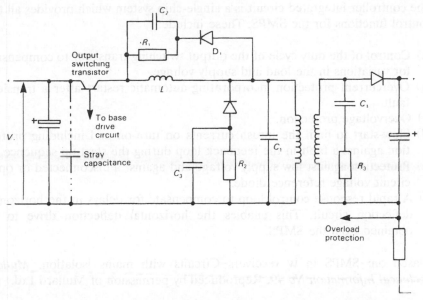

Fig. 5.11 Television receiver SMPS, control circuits to maintain output transistor within its SOA.

interference is the high di/dt of the current pulse produced by the output switching transistor.

(b) **Symmetrical mains pollution** – caused by the SMPS switching the current drawn through the main input rectifier.

(c) **Asymmetrical mains pollution** – appears between the earth and the supply line and neutral conductors. Its source is the capacitive currents flowing between the high level AC voltages and earth.

In addition to the measures already described for the limitation of visible interference caused by RFI, further control can be achieved by:

(a) Limiting the dV/dt in the rectifiers in the output transformers secondaries by the connection of a capacitor in parallel with the rectifier diode.

(b) Decreasing the dV/dt in the base drive transistor by means of an RC network in the collector of the output switching transistor.

(c) Providing a low-inductance path from the common line on the mains isolated side of the supply to the receiver chassis.

The mains interference can be controlled by:

(a) The inclusion of filtering in the mains input lines.

(b) Shielding the transformer to prevent stray voltages.

(c) Connection of a small capacitor (< 4.7 nF) between the mains isolated and non-isolated common rails.

159

The controller integrated circuit is a single-chip system which provides all the control functions for the SMPS. These include:

(a) Control of the duty cycle of the output switching transistor to compensate for variations in the load and supply voltage.
(b) Overcurrent protection incorporating automatic restart after a transient fault.
(c) Overvoltage protection.
(d) Slow-start to limit the inrush currents on turn-on and including protection against a fault in the feedback loop during the start-up sequence.
(e) Protection against low supply voltage and against a disconnected or open circuit voltage reference diode.
(f) A rapid response control loop to compensate for delays in the horizontal deflection circuit. This enables the horizontal deflection drive to be obtained from the SMPS.

(Based on: SMPS in tv receivers–Circuits with mains isolation. *Mullard Technical Information No 49*. Reproduced by permission of Mullard Ltd.)

Series-resonant power supply

Nijhof, E.B.G. and Evans, H.W. (1981). Introduction to the series-resonant power supply. *Electronic components and Applications*, **4** (1).

As the value of L_2 would normally be chosen to be at least 10 times that of L_1 its effect can be ignored.

C_o should have a value at least twice that of C_1.

The basic configuration of a series-resonant power supply (SRPS) is shown in Fig. 5.12. Initially, with the GTO thyristor T turned off, capacitor C_1 is charged to the supply voltage. If GTO thyristor T is now turned on just long enough to discharge C_1, then, when it is turned off, the current i_1 will oscillate about zero with a frequency determined by the value of inductor L_1 together with capacitors C_1 and C_o in series such that

$$\omega = \frac{1}{\sqrt{(L_1 C_x)}}$$
(5.31)

where $C_x = \dfrac{C_1 C_o}{C_1 + C_o}$

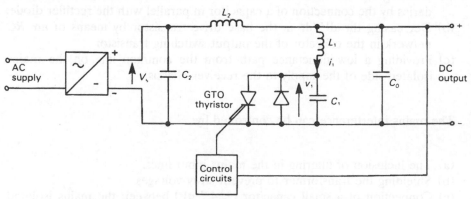

Fig. 5.12 Basic circuit of a series-resonant power supply.

For a self-stable oscillation the voltage v_1 across capacitor C_1 must reach zero in the course of each cycle. Voltage v_1 is defined by Equation 5.32 and varies between 0 and $2V_s$:

$$V_1 = V_s(1 - \cos \omega t) \tag{5.32}$$

The output voltage (v_o) is determined by the values of C_1 and C_o giving a peak value for v_o of

$$V_{o, \text{max}} = \frac{V_s(C_1 + C_o)}{C_o} \tag{5.33}$$

and it has an AC component of amplitude $\hat{V}_{o, \text{AC}}$ such that

$$\hat{V}_{o, \text{AC}} = \frac{V_s C_1}{C_o} \tag{5.34}$$

If the conduction period of the GTO thyristor T is increased, current will be flowing in inductor L_1 on turn-off, resulting in an increase in the peak oscillatory current by an amount N such that

$$N = 1 + \left(\frac{I_1 Z_1}{V_s}\right)^2 \tag{5.35}$$

N is the multiplying factor.

where Z_1 is the effective impedance of the L_1/C_1 branch.

This results in an increase in the AC component of v_o such that

$$\hat{V}_{o, \text{AC}} = \frac{N V_s C_1}{C_o} \tag{5.36}$$

This means that control of the output voltage can be achieved by varying the conduction period of the GTO thyristor T. Depending on the circuit conditions, multiplying factors (N) of 11 can be achieved and the output stabilized against input voltage variations of the order of 6 to 1.

Power can be taken from the SRPS in a variety of ways, some of which are shown in Fig. 5.13. In Fig. 5.13(a) the connection to the load is via a diode to a smoothing capacitor. Used in this way, the SRPS can be arranged to supply a stabilized output voltage with a small ripple at twice the oscillatory frequency. No smoothing is required on the input rectifier, reducing the distortion in the supply current, reducing the levels of harmonic current returned to the supply.

Chapter 7 gives a discussion of harmonics.

If instead, a capacitor-coupled output is used, an AC signal at the oscillatory frequency is available at the output. This can be either used directly or in association with other circuits such as the voltage doubler shown in Fig. 5.13(b).

In Fig. 5.13(c) the output is transformer-coupled with inductor L_2 incorporated into the transformer inductance. This enables the load to be matched to the SRPS by varying the turns ratio of the transformer. This configuration provides an isolated output and can be used to supply resistive or rectifier loads.

A further modification incorporates inductors L_1 within a transformer connection. In this form, the SRPS is inherently immune to a short circuit at its output and is self-starting under all conditions.

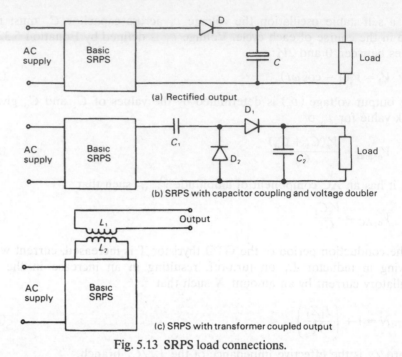

(a) Rectified output

(b) SRPS with capacitor coupling and voltage doubler

(c) SRPS with transformer coupled output

Fig. 5.13 SRPS load connections.

Uninterruptable power supplies

In applications such as hospital intensive care units, chemical plant process control, safety monitoring systems or major computer installations, even a temporary loss of supply could have serious consequences and there is therefore a requirement to maintain supply under all conditions. Initially, such uninterruptable power supply (UPS) systems were based on arrangements of the type shown in Fig. 5.14. Here, a DC motor, supplied from the AC mains via a converter, is used to drive a synchronous alternator which then provides the required AC supply. Also connected to the alternator shaft

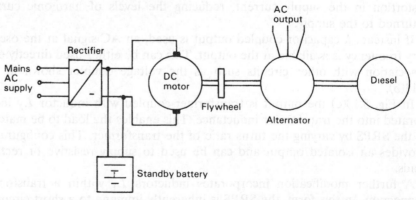

Fig. 5.14 Uninterruptable power supply system based on a DC motor–generator set with standby battery and diesel prime mover.

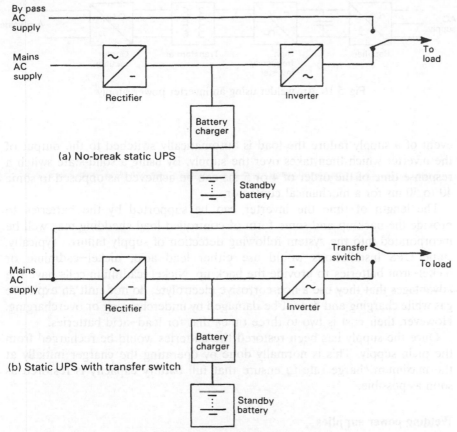

By pass AC supply

Mains AC supply

Rectifier

Battery charger

Inverter

To load

(a) No-break static UPS

Standby battery

Mains AC supply

Rectifier

Battery charger

Inverter

Transfer switch

To load

(b) Static UPS with transfer switch

Standby battery

Fig. 5.15 Static UPS configurations.

is a diesel engine which would start automatically following a failure of the main supply to and take over the load after a delay of some 10 to 15 s. By incorporating into the system an appropriate flywheel together with a standby battery supply to the DC motor generator speed could essentially be maintained during the start-up period of the diesel engine, ensuring a no-break supply at the alternator terminals.

Static UPS systems make use of power electronic switching devices and operate by rectifying the supply from the mains and using the resulting DC output to supply an inverter which then supplies the load as in Fig. 5.15. In the case of a failure of the AC supply, the supply to the inverter is first taken over by the battery bank, supplemented in some cases by a diesel engine to cover any long duration failure of the main AC supply. This arrangement is also used where a clean supply is required to condition the output voltage waveform and to protect the load from transients and harmonic contamination on the main AC supply. As the inverter is in continuous operation in the arrangement of Fig. 5.15(a), a bypass connection is provided to the AC supply in case of inverter failure.

Where a no-break supply is not essential the arrangements of Fig. 5.15(b) can be used. Here, the load is normally supplied by the main AC supply which also provides a trickle charge to top-up the standby batteries. In the

The battery charger would normally be arranged to provide a continuous trickle charging current to the standby batteries.

163

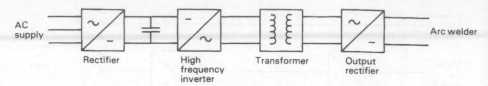

Fig. 5.16 Arc welder using an inverter power supply.

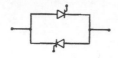

A solid state switch or relay typically consists of a pair of thyristors connected in reverse parallel. When the switch is ON they are fired at zero delay.

event of a supply failure the load is automatically switched to the output of the inverter which then takes over the supply. By using a solid-state switch a response time of the order of 4 or 5 ms can be achieved as opposed to some 40 to 50 ms for a mechanical contactor.

The length of time the inverter can be supported by the batteries to provide the back-up and some form of controlled load shedding may well be incorporated into the system following detection of supply failure. Typically, static UPS installations would use either lead–acid, nickel–cadmium or nickel–iron batteries to provide the back-up. Nickel–cadmium cells have the advantages that they use a non-corrosive electrolyte, do not emit an explosive gas while charging and cannot be damaged by undercharging or overcharging. However, their cost is two to three times that for lead–acid batteries.

Once the supply has been restored the batteries would be recharged from the main supply. This is normally done by operating the charger initially at the maximum charge rate to ensure that full battery capacity is restored as soon as possible.

Welding power supplies

A conventional DC arc welding system requires a high current at a voltage in the range 50 V to 100 V. The necessary step down is typically provided by a large heavy and expensive mains transformer which also provides the required isolation between the incoming supply and the welding arc. By increasing the transformer operating frequency, a smaller transformer can be used since for the same working flux density and output voltage a smaller core cross-sectional area is required at the higher frequency.

A welding supply operation on this principle is shown in Fig. 5.16. The incoming AC supply is first rectified and then used as the supply to the high-frequency inverter, the output of which is connected via a transformer and rectifier to supply the arc. By using asymmetric and GTO thyristors, supplies with ratings in excess of 15 kVA at frequencies of 30 kHz are achievable with efficiencies of the order of 80% to 85%.

Compared with an efficiency of 50% or less for a conventional welding supply.

Worked example 5.2

Estimate the change in core cross-sectional area required between ideal transformers of the same ratio and with the same maximum flux density in their cores when operating at frequencies of 5 Hz and 20 Hz.

Let N = Number of turns on primary (or secondary) of transformer
$\phi = \hat{\phi}\sin \omega t$

164

and $\quad\quad e_1 = N\dfrac{\mathrm{d}\phi}{\mathrm{d}t} = \omega N \hat{\phi} \cos \omega t$

Hence $\quad\quad E_{1,\,\mathrm{RMS}} = \dfrac{\omega N \hat{\phi}}{\sqrt{2}}$

For constant $E_{1,\,\mathrm{RMS}} \quad \hat{\phi} \propto \dfrac{1}{\omega}$

Now

$\quad\quad \hat{\phi} = \hat{B} \times (\text{Core cross-sectional area})$

where $\hat{B}$ is the maximum value of flux density in the core

Therefore, for constant $\hat{B}$

$\quad\quad$ Core cross-sectional area $\propto 1/\omega$

when

$$\dfrac{\text{Core cross-sectional area at 50 Hz}}{\text{Core cross-sectional area at 20 kHz}} = 400{:}1$$

Problems

5.1 A step-up converter used for an SMPS has to meet the following specification:

$V_o = 28\,\mathrm{V}$
$I_o = 50\,\mathrm{mA}$
$f_{\min} = 50\,\mathrm{kHz}$
$V_{\mathrm{ripple,\,p-p}} = 120\,\mathrm{mV}$

Estimate values for the peak in the transistor, the system inductance and the output capacitance for this circuit. Neglect the on state voltage drops of the transistor and diode.

5.2 What would be the effect on the values calculated in problem 5.1 of incorporating an on state voltage drop of 0.8 V for both the transistor and the diode? All other parameters remain the same.

6
Applications III

Objectives

□ To examine the application of power semiconductors to high-voltage DC transmission.
□ To examine the use of power semiconductors for voltage regulation.
□ To consider the operation of a thyristor circuit breaker.

High-voltage DC transmission

AC transmission dominates electricity supply by virtue of ease of generation, motor characteristics and the ability to change voltage magnitudes easily using transformers. High-voltage DC (HVDC) transmission uses a lighter, and hence cheaper, construction than an equivalently rated AC system with two conductors instead of three, and operation with cable is simplified because of the reduction of capacitive effects. HVDC systems do require complex and expensive terminal arrangements involving fully controlled converters. However, where bulk power is to be transmitted over long distance, either using overhead transmission, cables or some combination of these, for water crossings or as a connection between AC systems operating at different frequencies, HVDC may become economic for transmission distances of between 10 and 20 km.

Kimbark, E.W. (1971). *Direct Current Transmission: Volume 1.* John Wiley, New York.

HVDC systems are particularly viable for cable links, particularly involving water crossings and indeed the first modern HVDC system was the 96 km Gotland link installed in 1954 between Vostervik in Sweden and Visby on the island of Gotland. This was followed by other, similar links such as the original Cross-Channel link between England and France commissioned in 1961 and the Cook Strait link between the North and South Islands of New Zealand. A large number of such links are now operating worldwide with power ratings up to several gigawatts and voltages of $\pm$ 500 kV DC or higher.

A case study of the more recent Cross-Channel link is given later in this chapter.

Initially, development of HVDC systems was based on the use of mercury-arc valves with the first solid-state systems using thyristors being introduced in the early 1970s. From this point on, development has been based entirely on the use of thyristors. The result was that by 1983 the installed capacity in HVDC was almost equally divided between mercury-arc and thyristor based systems with all subsequent growth based on the latter.

Approximate installed capacity of HVDC worldwide:
1970– 5 GW
1980–14 GW
1990–35 GW

Figure 6.1 shows that the basic types of HVDC transmission system in use. The simplest arrangement is the monopolar link of Fig. 6.1(a) which uses a single conductor, usually at negative polarity, together with a ground or sea return. With the bipolar link of Fig. 6.1(b) uses both a positive and a negative conductor with the neutral point grounded at one or both ends of the system.

Negative polarity is preferred as it produces less interference.

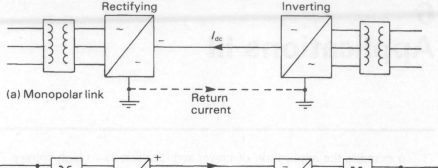

(a) Monopolar link

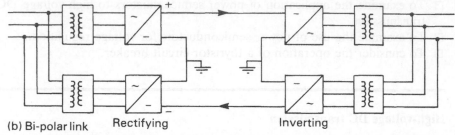

(b) Bi-polar link   Rectifying   Inverting

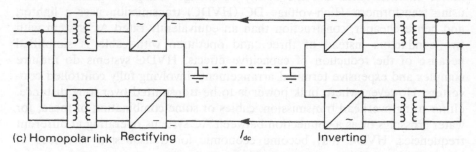

(c) Homopolar link   Rectifying I_{dc} Inverting

Fig. 6.1 Types of HVDC link.

Each terminal consists of a pair of converters connected in series on the DC side and in parallel on the AC side. Normal operation is with equal current in the positive and negative conductors in which case there is zero earth current. However, with both neutral points grounded, each of the poles can be operated independently with ground return, thus in the case of a fault on one pole, the other can still be used to supply a half-rated load. Finally, the homopolar link of Fig. 6.1(c) uses the same terminal connections as the bipolar link but both poles are operated at the same polarity, usually negative, with a ground or sea return.

HVDC converter

An HVDC converter will typically consist of a pair of six-pulse bridges connected in parallel on the AC side and in series on the DC side as in Fig. 6.2. Connection to the main AC supply will be by transformer, where by using a combination of a delta–delta transformer and a delta–star transformer to supply the bridges a 30° phase shift is introduced between the inputs to the

The term 'thyristor valve' is used in relation to HVDC systems to refer to the series/parallel grouping of individual thyristors used. The firing arrangements are such that externally this grouping looks and acts like a single large thyristor. See Chapter 1 for a discussion of the connection of thyristors in series and parallel.

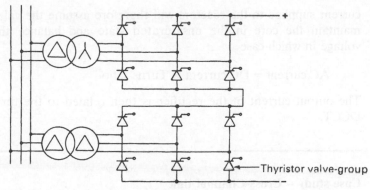

Fig. 6.2 HVDC 12-pulse converter formed by the connection of two six-pulse converters.

two bridges to produce effective 12 pulse operation as seen from the AC system. In practice, each of the **thyristor valves** shown in figure 6.2 is made up of a large number of individual thyristors connected in series/parallel to provide the required current and voltage ratings for the converter.

Typical control schemes for HVDC transmission systems include constant DC current, constant DC voltage, equidistant firing angle operation and, in the case of the inverting station, operation with constant extinction angle. Protection would also be incorporated into the control scheme. For example, in the case where a fault results in an increase in the current in the DC link, the firing angle of both the rectifying and inverting converters would be controlled to prevent the DC link current from exceeding a preset limit.

The DC current can be measured by means of Hall effect devices or the direct current current transformer (DCCT) of Fig. 6.3. Here, an AC voltage is applied to the secondary coils of the saturable reactors which are connected in series opposition. During any one half-cycle of the applied AC voltage the resulting flux will add to the DC flux in one core, causing it to saturate, and subtract from the corresponding flux in the other. There will then be no induced voltage in the coil associated with the saturated core since $d\phi/dt$ becomes zero during saturation, at the same time an ampere-turn balance will be set up in the coil associated with the unsaturated core. The AC

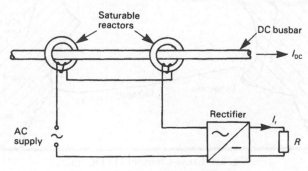

Fig. 6.3 DC current transformer.

current supplied to the reactors will therefore assume the value necessary to maintain the core in the unsaturated state and balance the applied AC voltage in which case

$$AC\ current = DC\ current \times Turns\ ratio \qquad (6.1)$$

The output current of the rectifier is then related to the turns ratio of the DCCT.

Case study – Cross-Channel link

(GEC Transmission and Distribution Projects Limited)

The first interconnection between the power systems of the Central Electricity Generating Board (CEGB) in the UK and Électricité de France was completed in 1961 with the commissioning of a 160 MW, ±100 kV HVDC link between converter stations using mercury-arc valves at Lydd and Echinghen (Fig. 6.4). The cables for the link were laid directly on the seabed and subsequently suffered damage from fishing trawls and anchors, and as a result were frequently out of service. The link was finally decommissioned in 1982.

The new link is rated at 2000 MV, ±270 kV and uses air-cooled thyristor valves in its converter stations at Sellindge near Ashford in Kent and Les Mandarins in France, including a 45 km submarine crossing of the Channel (Fig. 6.4). By burying the cables in trenches on the seabed damage should be minimized, enabling the design availability of 95% to be achieved.

Effectively two independent 1000 MW links.

The interconnection of the power systems of the UK and France has a number of features. Firstly, there is an increased diversity of availability of generating plant, with plant in one country available to support the other.

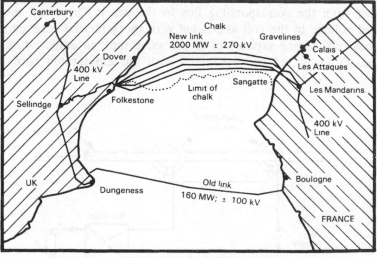

Fig. 6.4 Cross-Channel HVDC links.

The different patterns of demand can also be exploited, leading to more efficient plant operation while the country with higher marginal generating costs in any period can buy and import power from the other at lower cost.

The scheme

The Sellindge converter station occupies a 34-acre site on the route of the Dungeness/Canterbury 400 kV transmission line. From Sellindge the cables run to Folkestone and then across the Channel to the French coast at Sangatte and on to Les Mandarins.

In the UK, the land cables are of the oil pressure impregnated paper type, unarmoured with a conductor cross-section of 800 mm² and an overall diameter of 80 mm. The eight cables are laid in pairs in 1.0 m deep trenches.

The submarine cables are shared by the UK and France with each country providing four cables. The UK cables are of the mass impregnated solid type with a conductor cross-section of 900 mm² and an overall diameter of 105 mm. Electrical insulation is by means of 220 impregnated paper tapes wound helically over the stranded copper conductors. The cable is protected by lead and polythene sheaths and armoured with 5 mm steel wires, and is manufactured in 4 × 5 km continuous lengths.

The cables are laid as four independent pairs spaced 1 km apart on the seabed. Each pair consists of a positive and a negative cable laid in a 1.5 m deep trench cut in the chalk of the seabed. The pairs are laid touching in the trench to eliminate magnetic field effects which could interfere with ships' compasses.

The UK cables are laid in a two-stage process. In the first stage an unmanned, tracked, self-propelled trenching machine is used to cut a 1.5 m deep by 0.6 m wide trench. At the same time a guide hawser is laid in the trench by the machine. Once the trench is completed, the cable laying and embedding machine (CLEM) can begin installing the cables. This machine, operated from the cable laying vessel, uses the previously installed guide cable to pull itself along the trench, clearing the way with high pressure water jets. The two cables are fed down to the CLEM which positions them in the trench, at the same time the guide hawser is recovered to the surface. Backfilling of the trench is then by natural movement of the bottom sediment.

The converter station

Each country assumed the responsibility for the design and construction of its own converter station, with GEC Transmission and Distribution Projects the main electrical contractor for the UK terminal at Sellindge. The general circuit for Sellindge is shown in Fig. 6.5. Because of the space restrictions on the site, metal-clad SF$_6$ switchgear is used for the 400 kV substation. Each bipole of the HVDC link is configured to be switched separately at 400 kV, as is the reactive compensation and auxiliary supplies. The harmonic filters are switched at the bipole 400 kV busbars.

The 12 thyristor valves that make up each 500 MW pole use air-cooled thyristors and are arranged in three, four-unit combinations referred to as quadrivalves. Each thyristor valve consists of 125 modules connected in

Between 150 and 200 acres would have been required for an equivalent, conventional power station.

The RTM III trenching machine was developed by Land and Marine Engineering.

The CLEM was developed by Balfour Kirkpatrick.

The term 'thyristor valve' is defined on p. 168. The converter station consists of a total of 48 thyristor valves arranged to form 12 quadrivalves.

171

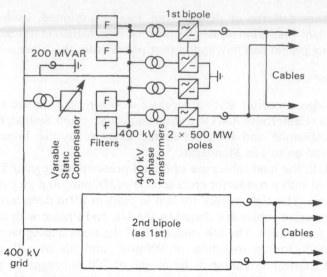

Fig. 6.5 Simplified circuit of Sellindge converter station.

Details of the capsule construction of thyristors are shown in Fig. 1.20(b).

Chapter 7.

series, a total of 6000 modules for the complete station. The modules themselves comprise a pair of parallel-connected thyristors 56 mm in diameter together with the components for the control of current and voltage stresses. The thyristors have a forward voltage capability of 3.2 kV and a reverse breakdown voltage of 4 kV. The current rating of the pair is 2000 A when operated in a full-wave bridge configuration. The nominal bridge rating is 135 kV and 1850 A using 125 series-connected modules.

The thyristors are of an established design using an inert-gas-filled, hermetically sealed capsule construction with double-sided cooling. An amplifying gate is incorporated to minimize gate power requirements. The gate signals for the valves are produced by a 12-pulse, equidistant-firing control system and transmitted to the module using fibre optic light guides.

When operating at full rated power each thyristor valve will have a loss of less than 200 kW, giving a total loss below 4.5 MW for each bipole. Each module has its pair of thyristors mounted on copper heat sinks and these, together with the main power connections, provide the outward heat transfer. The heat output from the valves is removed by a closed-cycle, filtered air system designed to provide cooling air at an input temperature of 40° C. The heated air is cooled by an air-to-water heat exchanger with final rejection to atmosphere by spray-type cooling towers.

Instead of the conventional series-tuned filters for the control of characteristic harmonics there is a series of eight filters which will be switched in combinations determined by loading. Second-order filters will be used to control characteristic harmonics with third-order filters introduced to improve the intrinsic damping of the AC system at low frequencies and to limit non-characteristic harmonics.

The Sellindge converter terminal requires some dynamic compensation to provide both reactive power absorption and generation to maintain operation at or near to unity power factor. A reactive overload absorption capability is also required to contain temporary overvoltages that might occur when the link is blocked, for example following a fault, before the filters are disconnected. Two high-speed compensators are installed at Sellindge. These are of the saturated reactor type and together with switched shunt capacitance provide ± 300 MVAR of compensation. In addition, during overloads the saturated reactors can each absorb 495 MVAR for 0.5 s. A third, identical compensator is installed on the 400 kV system at Ninfield near Hastings to provide support to the system west of Dungeness.

Control

The link can be controlled from either of the converter stations or, more usually, the grid control centres in the UK or France. Communication between the converter stations is by power line carrier over the eight DC cables. Converter control signals are transmitted at 1200 baud and instructions and signalling at 300 baud.

(Based on: Yates, J.B. and Arnold, R. (1985). Demand patterns make submarine cross channel link economic. *Modern Power Systems*, February.)

Goddard, S.C., Yates, J.B., Urwin, R.J., le Du, A., Marechal, P. and Michel, R. (1980). The new 2000 MW interconnection between France and the United Kingdom. *International Conference on Large High Voltage Electric Systems*, CIGRE, Paper 14–09.

Voltage regulation

Phase angle control

By connecting a reverse parallel pair of thyristors or a triac in each phase of the AC supply, the effective voltage applied to the load can be controlled though at the expense of increased harmonic content in both the load and the AC system. Figure 6.6 shows a single-phase voltage regulator operating with both resistive and inductive loads. For the resistive load:

$$V_{\mathrm{L,RMS}} = \left[\frac{1}{\pi} \int_{\alpha}^{\pi} (\hat{V} \sin\omega t)^2 \, \mathrm{d}(\omega t) \right]^{1/2}$$

$$= \hat{V} \left[(\pi - \alpha + \tfrac{1}{2}\sin 2\alpha)/2\pi \right]^{1/2} \tag{6.2}$$

This gives a mean power in the load of

$$P_{\mathrm{mean}} = \hat{V}^2(\pi - \alpha + \tfrac{1}{2}\sin 2\alpha)/2\pi R \tag{6.3}$$

With an inductive load the minimum possible firing angle for control corresponds to the phase angle of the load. For firing angles below this value, conduction will be continuous.

Three-phase loads can be controlled by either of the circuits shown in Fig. 6.7. For the fully-controlled circuit of Fig. 6.7(a), two thyristors must always be conducting. In order to ensure that this condition exists, the thyristors must receive a second gate pulse 60° after the initial pulse. This also covers those conditions where the current is discontinuous. The actual load current and voltage waveforms are complex and depend upon the number of devices conducting at any instant and the load impedance.

If one device in each of the AC supply lines is conducting then operation is as for a normal three-phase supply. When only two devices are conducting then the load is effectively receiving a single-phase supply and the third terminal assumes a potential which is the mean of the voltages on the two conducting phases.

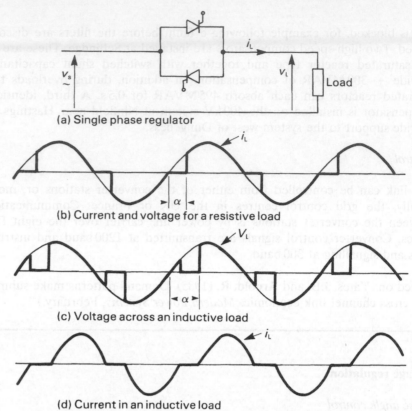

(a) Single phase regulator

(b) Current and voltage for a resistive load

(c) Voltage across an inductive load

(d) Current in an inductive load

Fig. 6.6 Single-phase regulator with resistive and inductive loads.

The half-controlled arrangement of Fig. 6.7(b) is simpler and requires only a single gate pulse as the return path is always provided by a diode. This results in a simplified control system and a cheaper installation.

Applications include incandescent lighting and heating.

Integral cycle control

Also known as burst firing.

A resistive load such as heating can be controlled by any of the circuits of Figs 6.6 and 6.7. Alternatively, where the thermal time constant of the load is sufficiently long, the load power can be controlled by switching the supply to the load on for an integral number of cycles of half-cycles and then off for a further integral number of cycles or half-cycles as in Fig. 6.8. The power delivered to the load is then

$$P_{\text{load}} = \frac{V_{\text{load}}^2}{R} \frac{N}{N+M} \qquad (6.4)$$

where N is the number of half-cycles during which the supply is connected to the load and M the number of half-cycles over which the supply is disconnected from the load.

Worked example 6.1

A single-phase resistive heating load is to be controlled from a single-phase, 50 Hz AC supply by means of an inverse parallel pair of thyristors. What will

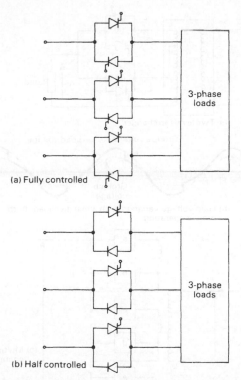

(a) Fully controlled

3-phase
loads

(b) Half controlled

3-phase
loads

Fig. 6.7 Regulators supplying three-phase loads.

be the firing angle of the thyristors when the load is 60% of its maximum value? If this controller is replaced by a burst firing system operating with a constant repetition period of 0.24 s, what will be the range of output powers available?

From Equation 6.2,

$$P_{max} = V_{max}{}^2/(2R)$$

Hence

$$\frac{P_{load}}{P_{max}} = (\pi - \alpha + \tfrac{1}{2}\sin 2\alpha)/\pi$$

When $P_{load} = 0.6P_{max}$

$$0.6 = (\pi - \alpha + \tfrac{1}{2}\sin 2\alpha)/\pi$$

Solving numerically gives

Firing angle $\alpha = 80°\ 55'$

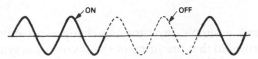

Fig. 6.8 Integral cycle control.

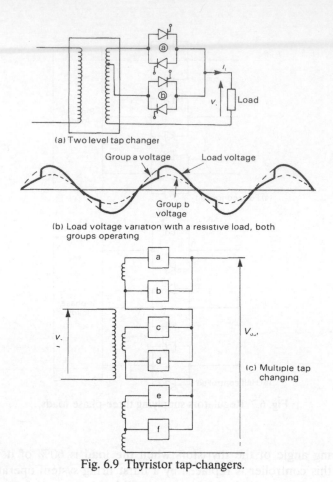

(a) Two level tap changer

(b) Load voltage variation with a resistive load, both groups operating

(c) Multiple tap changing

Fig. 6.9 Thyristor tap-changers.

For burst firing, the repetition frequency of 0.24 s represents 12 cycles at 50 Hz or 24 half-cycles.

Therefore, referring to Equation 6.4,

$$N + M = 24$$

and

$$\frac{P_{\text{load}}}{P_{\text{max}}} = N/(N + M) = N/24$$

Available powers range from 4.17% of P_{max} ($N = 1$) to 100% of P_{max} ($N = 24$), varying in steps of 4.17% of P_{max}.

Tap-changers

Figure 6.9(a) shows a simple thyristor tap-changer providing two levels of output voltage. Provided that the firing of the thyristors is synchronized to the voltage and current this circuit will not reduce the power factor or generate any harmonics. By varying the firing angle of the thyristors between the two pairs, additional control of the output voltage can be achieved as shown in Fig. 6.9(b). Where a greater range of variation in the output voltage is required then the arrangement of Fig. 6.9(c) could be used. By grading the secondary winding voltages and switching them in combination, a wide range of output voltages can be achieved.

Bird, B.M. and King, K.G. (1983). *An Introduction to Power Electronics*. John Wiley, U.K.

Thyristor circuit breakers

Currently, DC high-speed circuit breakers (HSCBs) are used for the protection of systems such as railway DC supplies and high-voltage DC motors. HSCBs are mechanical devices and suffer from a number of disadvantages. In particular, contacts are eroded by the switching of load currents and operation against faults, creating maintenance problems, operating speed may be inadequate to provide proper protection to semiconductor devices and the high temperature of the arcs on fault interruption creates problems of heat removal. The thyristor DC circuit breaker is a method of providing a fast, static means of interrupting DC current which is capable of repeated operation with low maintenance.

The principle of operation of the thyristor DC circuit breaker is shown by Fig. 6.10(a). In normal operation the current path is as in Fig. 6.10. When the fault occurs the rise in the current through the main thyristor is detected and the auxiliary thyristor turned on, discharging the capacitor through the main thyristor to commutate it off. This condition is shown in Fig. 6.10(b). The current path is now via the auxiliary thyristor as in Fig. 6.10(c) and the capacitor charges in the reverse direction causing the auxiliary thyristor to turn-off when the current through it falls below its holding level. Energy in the load inductance is then dissipated by the circulating current in the freewheeling diode as in Fig. 6.10(d).

Forced commutation circuits are described in Chapter 3.

Table 6.1 sets out some of the stages in the development of thyristor circuit breakers. The majority of thyristor circuit breaker applications to date have been on railway traction systems where their capacity for repeated operation, low maintenance costs and the ability to dissipate the energy in the catenary system offset the increase in cost relative to mechanical HSCBs.

Contactors

An electronic or solid-state contactor or relay performs the same circuit function as a mechanical contactor or relay in providing isolation of a load from the supply. Mechanical contactors and relays are relatively slow with operating times of the order of a few tens of milliseconds and this can cause problems in applications such as power supplies or reversing drives.

A solid-state contactor typically consists of a pair of thyristors connected in reverse parallel as in Fig. 6.11(a) or a triac as in Fig. 6.11(b). In operation, a firing pulse is applied to the gate at voltage zero to ensure continuous conduction. Turn-off will then be at the current zero following the removal of the firing pulse.

A snubber circuit would normally be incorporated across the solid-state contactor to accommodate switching transients.

As a solid-state relay does not provide for the physical isolation of the supply from the load, an isolating switch may be incorporated in series with

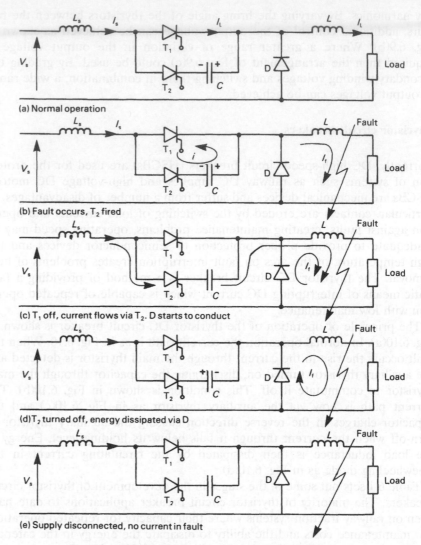

(a) Normal operation

(b) Fault occurs, T$_2$ fired

(c) T$_1$ off, current flows via T$_2$. D starts to conduct

(d) T$_2$ turned off, energy dissipated via D

(e) Supply disconnected, no current in fault

Fig. 6.10 Thyristor circuit breaker operation.

the load. This isolating switch is not rated to make or break load current but is operated only under conditions of zero load current.

The advantages of solid state contactors include speed and consistency of response, no routine maintenance requirement and no audible noise on switching.

Microelectronics and power electronics

Microprocessors and microelectronics are playing a significant and expanding role in the real-time control of power electronic systems ranging from variable-speed drives to power transmission, resulting in increased performance and efficiency by optimization of system operation. Where a single processor is used for this purpose, then its interrupt structure becomes

Table 6.1 Development of Thyristor Circuit-breakers
(Prof. P. McEwan, personal communication)

Thyristor circuit-breaker	Rated voltage (V)	Rated current (A)	Trip setting (A)	Peak let-through current (A)	Operating time (ms)	Application
Goldberg (1963)	120	5	12	25	0.04	Inverse time overload and short-circuit protection.
Zyborski (1976)	600	800	500 to 1500	1850	1.02	Overcurrent protection. Gdansk trolley-bus system.
Zyborski (1980)	1000		2700	3400	4	Experimental switch for overcurrent protection.
Modified Mazda (1982)	110	80	100		1.4	Experimental switch for low voltage DC with auto reclose.
Mitsubishi (1982)	1500	1300	4500	9700	< 10	Kotoni traction system, Sapporo.
Meidenisha (1982)	1500		3000	4200		Tozai subway line, Sapporo
Toshiba (1982)	1500	3000	6000	7500	2.9	Teito Rapid Transit Authority

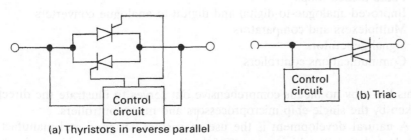

(a) Thyristors in reverse parallel   (b) Triac

Control circuit

Control circuit

Fig. 6.11 Solid-state contactor.

important along with the assignment of priorities within the system. However, with the reduction in the cost of microprocessors there is an increasing tendency towards the distribution of intelligence throughout a system with individual microprocessors allocated to the control of specific functions within the overall system. These developments, together with the ability of the microprocessor to handle complex algorithms, have increased operational flexibility and enabled system characteristics to be more closely matched to load requirements. These conditions are of particular importance where a number of functions such as the individual joint movements of a robot or the drive rollers of a paper mill need to be co-ordinated.

In applications where speed of operation is a significant factor the programs for the microprocessor would be written in the assembly language of the particular processor. Such an approach, while providing programs which are fast and of minimum size, is expensive in development time. By writing in a high-level language, software costs are reduced at the expense of a less efficient program. As a compromise, the bulk of the program may be written in the high-level language, using the assembly language only for the time-critical areas.

Since the introduction of the first microprocessors the developments in manufacturing technology have resulted in devices with architectures ranging from 4 to 32 bits. These are supported by an increasing range of single-chip microprocessors and microcontrollers. For applications involving high production volumes these single-chip devices provide a cost-effective approach to hardware.

Single-chip microcontrollers have been developed specifically for control applications and incorporate timers, counters, analogue-to-digital converters, input/output structures and RAM and ROM memory on the same chip. Future developments are likely to involve the incorporation of some of the following features on to the single-chip device:

Improved processor architecture
Increased on-board RAM and ROM
Enhanced input/output (I/O) facilities
Memory management units
Hardware multiply/divide
Enhanced timers
Watchdog timers
Phase-locked loops
Improved analogue-to-digital and digital-to-analogue converters
Multiplexers and comparators
Signal generators
Communications controllers

This list is by no means comprehensive but serves to illustrate the directions taken by the single-chip microprocessors and microcontrollers.

A natural development is the use of custom-designed and manufactured integrated circuits taking account of the improvements in device technologies and manufacturing techniques. When used in combination with power semiconductors such circuits form the basis of 'smart power' devices. These will carry on board the necessary I/O for both control and status information, isolation, required intelligence and logic, protection, packaged power supplies and device drive circuits. Manufacture could either be as a monolithic integration with all functions incorporated on a single chip or in a hybrid form with separate logic and power chips.

See also Chapter 1.

Applications for custom integrated circuits include the production of the switching sequence for stepper motors or the generation of inverter PWM waveforms for AC motor control. Smart power devices would include intelligent switches capable of monitoring and reporting their own status, incorporation within the housing of a stepper motor to control operation or, with appropriate communications included, a part of a network structure integrating a number of systems.

Case study – microelectronic motor speed controller

Series universal motors up to 750 W (1 HP) rating and running at speeds up to 12 000 rev min^{-1} are used in a wide range of domestic appliances such as automatic washing machines, duplicators, food mixers, portable drilling machines and saw benches. The requirements for the associated motor controller are low cost, high reliability and ease of application. The use of specially designed integrated circuits reduces the number of components, leading to a more reliable unit.

The basic speed-control module can have many derivatives as the basic circuit design can be modified readily to suit individual needs such as the provision of current limit control and additional speed settings. For example, the module can be configured to provide a wide range of speed settings over a range of system voltages.

The series universal or series commutator motor is a DC series motor operating from an AC supply. The action of the commutator means that the field and armature currents reverse direction together, maintaining a constant direction for torque. Motors of this type are capable of operating at high speeds.

Various forms of protection are incorporated within the module. In the case of a tachogenerator failure, either open or short circuit, the signal from the tachogenerator is lost and the module brings the motor to zero speed rather than allowing any acceleration to a possibly hazardous condition. Should the motor become stalled for any reason the module will again shut down the power supply to prevent burn-out. Reaction in either of the above two conditions is within 1 s.

The module is protected against the failure of the motor triac by the use of a discrete, low-cost, sacrificial transistor which is used to drive the triac and protect the more expensive integrated circuit. Following a failure of this type the module can easily and cheaply be repaired for subsequent re-use. The drive triac is rated at 12 A at 500 V, will take an inrush current of 85 A and can withstand a stalled rotor current of 30 A for 4 s.

A necessary feature for the intended applications in domestic appliances is protection against the voltage spikes associated with series commutator motors. Immunity is therefore provided against mains spikes up to 2000 V.

Features of particular interest are the provision of a soft-start facility, of particularly importance with belt drives to ensure that there is no belt slippage while accelerating in order to maximize belt life, and a programmed restart facility following power failure.

Frequency sensing is by means of a tachogenerator on the motor shaft. This enables factory presetting for multiple speed ranges such as might be used by a washing machine wash-and-spin cycle.

Problems

6.1 A bipolar HVDC transmission system is rated at 2000 MW, ± 320 kV and uses the converter arrangement of Fig. 6.2. Find the RMS current and peak reverse voltage for each of the thyristor valves.

6.2 A three-phase resistive heating load is controlled by triacs from a 415 V (line), 50 Hz supply. If the maximum load is 24 kW, determine the rating of the triacs and their firing angles for loads of 16 kW and 8 kW. If the triacs are replaced by thyristors, how would the current ratings change?

6.3 A single-phase load of resistance 12 Ω in series with an inductance of 24 mH is fed from a 240 V (RMS), 50 Hz supply by a pair of inverse parallel thyristors. Find the mean power in the load at firing angles of (a) 0°, (b) 90° and (c) 120°. Ignore source inductance and device voltage drops.

6.4 A tap-changer such as that shown in Fig. 6.9 is used to supply a resistive load. If the lower tapping is at 67% of full voltage, plot a curve showing the variation of the RMS voltage of the load against the firing angle of the 100% tapping thyristors.

7

Harmonics and interference

Objectives

- [ ] To consider the representation of harmonic sources.
- [ ] To consider briefly the effect of harmonics on system components.
- [ ] To examine AC and DC system harmonics.
- [ ] To consider the harmonics produced by inverters and integral cycle controllers.
- [ ] To examine the basic techniques of harmonic filtering and to introduce filter types.

Converters operate by switching the AC supply to produce a DC voltage at their output. The resulting deviation of the current drawn from the AC system from the ideal sinusoidal waveform results in the introduction of harmonic currents and voltages into both the AC and DC systems. Similarly, inverters switch a DC supply to produce an alternating, but non-sinusoidal, output which contains a significant harmonic content. The presence of these harmonic currents and voltages can influence system performance in a number of ways. Among these are:

Arrillaga, J., Bradley, D.A. and Bodger, P.S. (1985). *Power System Harmonics*. John Wiley, U.K.

(a) Amplification of system current and voltage levels as a result of series or parallel resonances with possible damage to system components due to overcurrents or overvoltages.
(b) Increased losses in items of system components such as transformers, motors and generating plant necessitating the derating of those components.
(c) Ageing of insulation as a result of overvoltages leading to reduced operational life.
(d) System maloperation, particularly computers and protection.
(e) Interference with communications systems.

Converter harmonics

Figure 7.1 shows a general, p-pulse, fully-controlled converter together with the quasi-square waveform representing the current waveform for one phase of the AC supply. The Fourier series ($F(t)$) for this waveform is

Fourier analysis is discussed in Appendix B.

$$F(t) = \frac{4}{\pi}\left[\sin\frac{\psi}{2}\cos(\omega t) + \frac{1}{3}\sin\frac{3\psi}{2}\cos(3\omega t)\right.$$

$$\left. + \frac{1}{5}\sin\frac{5\psi}{2}\cos(5\omega t) + ...\right] \tag{7.1}$$

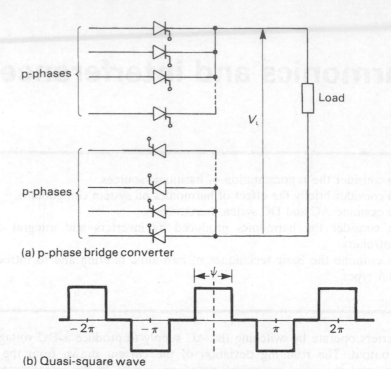

(a) p-phase bridge converter

(b) Quasi-square wave

Fig. 7.1 Operation of a general *p*-pulse bridge.

For an ideal six-pulse converter $\psi = 2\pi/3$ when, for a DC current amplitude of I_{DC}, the current in phase 'a' is

$$i_a = \frac{2\sqrt{3}}{\pi} I_{DC}\left[\cos \omega t - \frac{1}{5}\cos (5\omega t) + \frac{1}{7}\cos(7\omega t)\right.$$

$$- \frac{1}{11}\cos (11\omega t) + \frac{1}{13}\cos (13\omega t) - \frac{1}{17}\cos (17\omega t)$$

$$\left. + \frac{1}{19}\cos (19\omega t) - ... + ... - ...\right] \tag{7.2}$$

This contains harmonics of the order $n = 6k \pm 1$, where $k = 1, 2, 3$, etc., decreasing in magnitude with respect to the fundamental by

$$I_n = I_1/n \tag{7.3}$$

where *n* is the harmonic number.

Similarly, the phase current for a 12-pulse converter can be obtained as

$$i_a = \frac{4\sqrt{3}}{\pi} I_{DC}\left[\cos \omega t - \frac{1}{11}\cos (11\omega t) + \frac{1}{13}\cos (13\omega t)\right.$$

$$\left. - \frac{1}{23}\cos (2\omega t) + \frac{1}{25}\cos (25\omega t) - ... + ...\right] \tag{7.4}$$

This contains harmonics of the order $n = 12k \pm 1$, decreasing in magnitude according to Equation 7.3. From Equations 7.2 and 7.4 the supply system harmonic components of an ideal *p*-pulse converter are seen to be at orders of

$pk \pm 1$, where *p* is the pulse number and $k = 1, 2, 3$, *etc.*

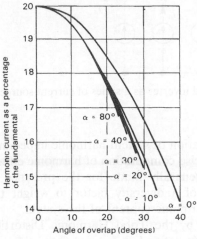

(a) Variation of 5th harmonic current in relation to firing angle and overlap angle

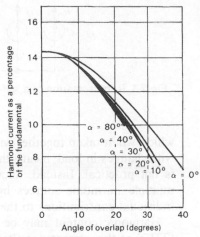

(b) Variation of 7th harmonic current in relation to firing angle and overlap angle

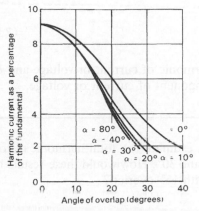

(c) Variation of 11th harmonic current in relation to firing angle and overlap angle

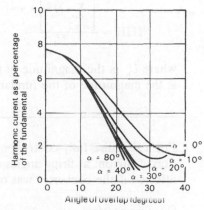

(d) Variation of 13th harmonic current in relation to firing angle and overlap angle

Fig. 7.2 Effect of firing angle and overlap on harmonic content.

However, a real converter operates with overlap and this will cause the actual harmonic content to vary from these ideal values. Figure 7.2 shows the effect of varying both the overlap and firing angle on the system harmonic currents for various harmonics.

In terms of the AC system, and as suggested by Fig. 7.3, a self-commutated converter can be considered as consisting of a series of current generators at the individual harmonic frequencies contained in the AC current waveform. These harmonic currents then combine with the system harmonic impedances to produce the harmonic voltages.

Where more than one harmonic source is present, the net effect on the power system is not obtained by the simple summation of the individual harmonic components and the phase angle relationship of the different sources must be taken into account. In many power systems, the distribution of the phase angles of the individual harmonic sources is essentially random

The curves of Fig. 7.2 are sometimes referred to as Read's curves.

The principle of superposition can be applied to the individual harmonics.

185

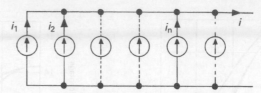

Fig. 7.3 Representation of a self-commutated inverter as a series of current sources.

which, when taken together with the variation of system harmonic impedance with changes in load, means that the precise computation of harmonic effects is not practical. Instead, allowance is generally made for the presence of multiple harmonic sources by the use of a diversity factor to weight the individual contributions to the overall harmonic content of the system.

Harmonic content may be expressed by the Total Harmonic Distortion (THD) of the current or voltage waveform defined as

$$\text{THD} = \frac{\left[\sum_{n=2}^{\infty} U_n^2\right]^{1/2}}{U_1} \tag{7.5}$$

where U_n is the magnitude of the nth harmonic of current or voltage and U_1 is the magnitude of the fundamental component of current or voltage.

Worked example 7.1

For a six-pulse fully-controlled converter, estimate the total harmonic distortion of current at firing angles of 0°, 10° and 30°. How would these results be affected if the converter was operating with 15° overlap?

(1) *Zero overlap*

$\alpha = 0°$; amplitude of fundamental current = 100%

From the curves of Fig. 7.2:

Amplitude of 5th harmonic $= I_5 = 20\%$
Amplitude of 7th harmonic $= I_7 = 14.3\%$
Amplitude of 11th harmonic $= I_{11} = 9.1\%$
Amplitude of 13th harmonic $= I_{13} = 7.7\%$

$$\text{Total harmonic distortion (THD) of current} = \frac{\left[\sum_{n=2}^{\infty} I_n^2\right]^{1/2}}{I_1}$$

$$= (20^2 + 14.3^2 + 9.1^2 + 7.7^2)^{1/2} = 27.3\%$$

(30% if all harmonics to 50th are included)
This result is the same for all firing angles.

186

(2) 15° *overlap*
(i) $\alpha = 0°$; amplitude of fundamental current = 100%
From curves:

$I_5 = 19.4\%$ $I_{11} = 7.5\%$
$I_7 = 13.3\%$ $I_{13} = 5.6\%$
THD of current = 25.3%

(ii) $\alpha = 10°$; amplitude of fundamental current = 100%
From curves:

$I_5 = 18.8\%$ $I_{11} = 6.5\%$
$I_7 = 12.8\%$ $I_{13} = 5\%$
THD of current = 24.2%

(iii) $\alpha = 30°$; amplitude of fundamental current = 100%
From curves:

$I_5 = 18.8\%$ $I_{11} = 6.5\%$
$I_7 = 12.7\%$ $I_{13} = 4.8\%$
THD of current = 24.1%

Where there is insufficient inductance in the load, the AC system currents may contain a significant ripple as in Fig. 7.4(a). This ripple can be related to the associated ripple on the DC current waveform when, referring to Fig. 7.4(b), expressions can be obtained for the fundamental component and the characteristic harmonics in terms of the ripple factor r, where

$$r = I_r / I_{DC} \qquad (7.6)$$

where I_r is the peak-to-peak amplitude of the ripple current and I_{DC} is the mean DC current

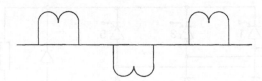

Fig. 7.4a Converter currents including ripple.

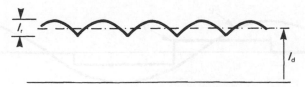

Fig. 7.4b AC current waveform.

Dobinson, L.G. (1975). Closer accord on harmonics. *Electronics and Power*, May, 567–72.

The fundamental component of current is given by

$$I_1 = I_d(1.102 + 0.014r) \tag{7.7}$$

The harmonic currents, expressed as a percentage of the fundamental are then

$$I_n = 100\left(\frac{1}{n} + \frac{6.46r}{n-1} - \frac{7.13r}{n}\right)(-1)^k \tag{7.8}$$

for $n = kp - 1$

and

$$1_n = 100\left(\frac{1}{n} + \frac{6.46r}{n+1} - \frac{7.13r}{n}\right)(-1)^k \tag{7.9}$$

for $n = kp + 1$

Full output voltage is obtained for a firing angle of $0°$ ($\alpha = 0°$).

Where regeneration is not a requirement, half-controlled bridges provide a cheaper solution than the fully-controlled bridge. When operated at full voltage ($\alpha = 0°$), the harmonic currents produced by a half-controlled bridge correspond to those of the fully-controlled bridge at full voltage. However, as firing angle is increased the current waveform loses its half-wave symmetry, as is shown by Fig. 7.5. Under these conditions the half-controlled bridge can generate high levels of harmonic current and under very light load conditions a second harmonic component can result whose amplitude approaches that of the fundamental.

Capacitor smoothing

Capacitor smoothing is used extensively with inverters operating from a single-phase AC supply.

As an alternative to relying on load inductance, converters can be operated with capacitive smoothing on their output. The resulting voltage and current waveforms are then as shown in Fig. 7.6 for a single-phase converter from

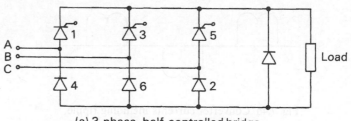

(a) 3-phase, half-controlled bridge

(b) Supply current and voltage waveforms, $\alpha = 60°$

Fig. 7.5 Operation of a three-phase half-controlled bridge.

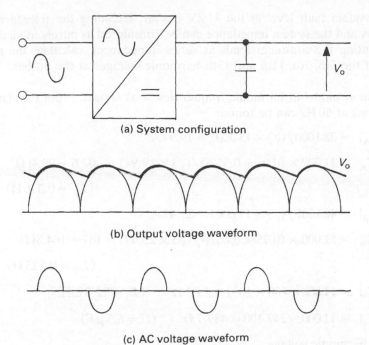

(a) System configuration

(b) Output voltage waveform

(c) AC voltage waveform

Fig. 7.6 Converter with capacitor smoothing.

which it can be seen that current is drawn from the AC supply over a relatively short part of each half cycle, resulting in high levels of harmonics being introduced into the supply.

Worked example 7.2

In the system shown, consumer A has a connected load of 380 kVA at a power factor of 0.95 lagging and wishes to connect a variable-speed AC drive using an uncontrolled six-pulse converter at the 11 kV busbar. The full-load fundamental current of the converter is measured at 4.2 A per phase and it operates with a 10° overlap angle. Consumer B has a normal connected load of 485.3 kVA at a power factor of 0.75 lagging and has installed 247.1 kVA of power factor correction capacitors at the busbar.

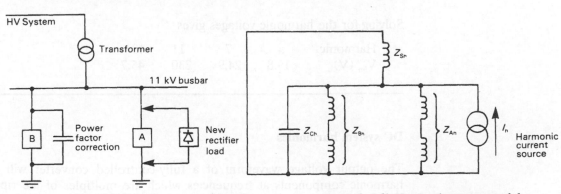

Example 7.2a System configuration.   Example 7.2b Harmonic system model.

The system fault level at the 11 kV busbar, including the transformer, is 30 MV A and the system impedance can be considered as purely inductive. By representing the consumer loads as series impedances, calculate the magnitudes of the 5th, 7th, 11th and 13th harmonic voltages at the busbar.

The system model at harmonic frequencies is as shown. From the data, the conditions at 50 Hz can be found:

$$|I_A| = 380\,000/(\sqrt{3} \times 11\,000) = 19.94\,\text{A}$$

$$Z_A = 11\,000 \times (0.95 + 0.312\text{j})/(\sqrt{3} \times 19.94) = 302.6 + 99.4\text{j}\,\Omega$$

$$(L_A = 0.316\,\text{H})$$

$$|I_B| = 485\,300/(\sqrt{3} \times 11\,000) = 25.47\,\text{A}$$

$$Z_B = 11\,000 \times (0.75 + 0.661\text{j})/(\sqrt{3} \times 25.47) = 187 + 164.8\text{j}\,\Omega$$

$$(L_B = 0.525\,\text{H})$$

$$|Z_s| = 11\,000^2/(30 \times 10^6) = 4.03\,\Omega \qquad (L_s = 12.8\,\text{mH})$$

$$|Z_c| = 11\,000^2/247\,100 = 489.7\,\Omega \qquad (C = 6.5\,\mu\text{F})$$

Busbar harmonic voltage:

$$V_h = Z_{eh} I_h$$

where
Z_{eh} is the effective harmonic impedance formed by the individual harmonic impedances Z_{Ah}, Z_{Bh}, Z_{Sh} and Z_{Ch} in parallel.
From above:

$$Z_{Ah} = 302.6 + 99.4n\text{j}\,\Omega \qquad Z_{Sh} = 4.03n\text{j}\,\Omega$$

$$Z_{Bh} = 187 + 164.8n\text{j}\,\Omega \qquad Z_{Ch} = -489.7\text{j}/n\,\Omega$$

From curves, the harmonic currents are

$I_5 = 19.8\%$ of $4.2\,\text{A} = 0.832\,\text{A}$
$I_7 = 13.8\%$ of $4.2\,\text{A} = 0.58\,\text{A}$
$I_{11} = 8.4\%$ of $4.2\,\text{A} = 0.353\,\text{A}$
$I_{13} = 6.8\%$ of $4.2\,\text{A} = 0.286\,\text{A}$

Solving for the harmonic voltages gives

Harmonic	5	7	11	13
V_{hn} (V)	19.8	24.9	230	45.7

DC system harmonics

The output voltage waveform of a fully-controlled converter will contain harmonic components at frequencies which are multiples of the ripple frequency. Figure 7.7 shows the DC voltage waveform for a six-pulse converter

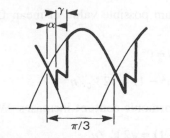

Fig. 7.7 Output voltage waveform for a six-pulse fully-controlled bridge.

supplying a constant DC current with a firing angle of α and an overlap angle of γ. This voltage can be represented by

$$v_d = \hat{V} \cos(\omega t + \pi/6) \qquad\qquad 0 < \omega t < \alpha \qquad (7.10)$$

$$v_d = \hat{V}[\cos(\omega t + \pi/6) + \cos(\omega t - \pi/6)]/2 \quad \alpha < \omega t < (\alpha + \gamma) \quad (7.11)$$

$$v_d = \hat{V} \cos(\omega t - \pi/6) \qquad\qquad (\alpha + \gamma) < \omega t < \pi/3 \qquad (7.12)$$

Using Fourier analysis the RMS magnitude of the nth harmonic component of the DC voltage waveform can be obtained in terms of the overlap angle γ, by Fourier analysis:

$$V_n = V_0\{(n-1)^2 \cos^2[(n+1)\gamma/2] + (n+1)^2 \cos^2[(n-1)\gamma/2]$$
$$- 2(n-1)(n+1) \cos[(n+1)\gamma/2] \cos[(n-1)\gamma/2]$$
$$\times \cos(2\alpha + \gamma)\}^{1/2} \, 1/[\sqrt{2}(n^2 - 1)] \qquad (7.13)$$

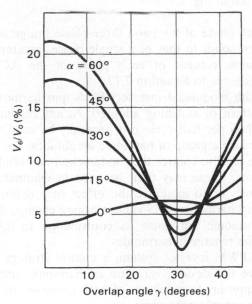

Fig. 7.8 Variation of sixth harmonic voltage content in the output of a six-pulse bridge converter.

$V_0 = 3\hat{V}/\pi$

Equation 2.21.

where V_0 is the maximum possible value of mean DC voltage and n is the harmonic number

With $\alpha = 0°$ and $\gamma = 0°$,

$$V_n = \sqrt{2}\,V_0/(n^2 - 1) \approx \sqrt{2}\,V_0/n^2 \tag{7.14}$$

As α is increased the harmonic content increases. With $\alpha = 90°$ and $\gamma = 0°$,

$$V_n = \sqrt{2}\,V_0 n/(n^2 - 1) \approx \sqrt{2}\,V_0/n \tag{7.15}$$

an increase of n times in the amplitude of the harmonic voltage. Figure 7.8 shows the relationship between harmonic voltage amplitude and overlap angle of Equation 7.13 plotted at various firing angles (α) for the sixth harmonic.

Inverter harmonics

For the single-phase inverter with a squarewave output the voltage contains odd harmonics only, such that the nth harmonic voltage is

$$V_n = V_1/n \tag{7.16}$$

where V_1 is the fundamental amplitude

If voltage control is applied to modify the output to quasi-squarewave form then, by reference to Fig. 7.1 and Equation 7.1, the amplitudes of the individual harmonics can be expressed in terms of the amplitude of the fundamental as

$$V_n = V_1 \sin(n\psi/2)/[n\sin(\psi/2)] \tag{7.17}$$

where ψ is as shown in Fig. 7.1

The output of each phase of the basic three-phase bridge inverter discussed in Chapter 3 corresponds to that of a single-phase inverter with $\psi = 2\pi/3$. Hence the harmonic content of each phase of the AC supply can be determined by reference to Equation 7.17.

Bird, B.M., King, K.G. and Pedder, D.A.G. (1993). *An Introduction to Power Electronics*. John Wiley, U.K.

Arrillaga, J., Bradley, D.A. and Bodger, P.S. (1985). *Power System Harmonics*. John Wiley, U.K.

For a pulse-width modulated inverter the output harmonics can be controlled by the pattern of switching adopted. At any fundamental switching frequency each chop per half cycle can generally be used to eliminate one harmonic or to reduce a group of harmonic amplitudes. For r chops per half cycle one must be used to control the fundamental amplitude leaving $(r - 1)$ degrees of freedom. These may then be used to eliminate $(r - 1)$ specific low-order harmonics or to minimize the effect of a defined range of harmonics. As the total harmonic RMS voltage cannot change, the elimination of any particular harmonic will cause its contribution to this voltage to be distributed over the remaining harmonics.

In designing a PWM inverter system, a control strategy must be chosen which will achieve the desired variation of harmonic amplitude with frequency with an appropriate reduction of both harmonic torques in a motor load and overall harmonic power loss.

A pulse-width-modulated inverter switches the supply in the following sequence in one half cycle:

(a) 26.63°, on; (b) 34.2°, off; (c) 53.95°, on; (d) 66.9°, off; (e) 82.61°, on; (f) 97.39°, off; (g) 113.1°, on; (h) 126.05°, off; (i) 145.8°, on; (j) 153.37°, off.

Find the amplitude of the harmonic components up to the 20th of this waveform in relation to the amplitude of the fundamental component.

Using the appropriate symmetry (Appendix 2) it is seen that the waveform contains only some terms and odd harmonics. In which case

$$b_n = \frac{4V_s}{\pi} \left(\int_{26.63°}^{34.2°} \sin n\theta \, d\theta + \int_{53.95°}^{66.9°} \sin n\theta \, d\theta + \int_{82.61°}^{90°} \sin n\theta \, d\theta \right)$$

$$= [\cos(26.63n) + \cos(53.95n) + \cos(82.61n) - \cos(34.2n)$$

$$- \cos(66.9n)$$

Solving,

$$b_1 = 0.5; \quad b_3 = 6.48 \times 10^{-4}; \quad b_5 = -5.23 \times 10^{-4}; \quad b_7 = -5.25 \times 10^{-4}$$

$$b_9 = = 4.45 \times 10^{-2}; \quad b_{11} = -0.361; \quad b_{13} = 0.361; \quad b_{15} = 4.31 \times 10^{-2};$$

$$b_{17} = 1.49 \times 10^{-3}; \quad b_{19} = 1.67 \times 10^{-2}$$

Integral cycle control

When using integral cycle control, the AC source frequency cannot be used as the fundamental frequency for Fourier analysis. Consider a controller with an ON period of N cycles, repeated every M cycles. The repetition period is then M/f seconds and the fundamental frequency, f_0, is f/M Hz. Referring to Fig. 7.9, in the interval $\omega_0 = 2\pi f_0$:

$$i_1 = I_m \sin(M\omega_0 t) \tag{7.18}$$

The waveform and period shown can be represented by a Fourier series.

The fundamental frequency is defined by the period over which the waveform repeats itself.

f is the frequency of the AC supply in hertz.

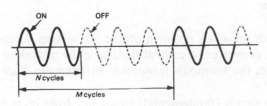

Fig. 7.9 Integral cycle control.

Therefore

$$b_n = \frac{2}{\pi} \int_0^{N\pi/M} [I_m \sin(M\omega_0 t) \sin n\omega_0 t] \, d(\omega t)$$

$$= (-1)^N I_m \, 2M \sin(Nn\pi/M)/[\pi(M^2 - n^2)] \qquad (7.19)$$

When $n = kM$, $k = 2, 3, 4, 5$, etc. then $b_n = 0$.

Integral cycle control of this type therefore produces no harmonics of the supply frequency but generates inter-harmonic and **sub-harmonic** frequency components.

Sub-harmonics are the principle source of flicker. This is often a problem where lighting circuits, particularly using fluorescent lights, are affected. Flicker is controlled by standards such as BS5406 and IEC555.

Worked example 7.4

An integral cycle controller is working from a 50 Hz supply. For the switching sequence of one cycle on followed by four cycles off, find the amplitudes of the first 20 harmonics.

The sequence repeats every five cycles, hence the fundamental frequency f_1 is $50/5 = 10$ Hz.

For the case given, $N = 1$ and $M = 5$ in Equation 7.19. Substituting in this equation for $n = 1$ to $n = 20$ gives

f_1	(10 Hz)	= 0.087	f_2	(20 Hz)	= 0.14
f_3	(30 Hz)	= 0.189	f_4	(40 Hz)	= 0.208
f_5	(50 Hz)*	= 0.2	f_6	(60 Hz)	= 0.17
f_7	(70 Hz)	= 0.126	f_8	(80 Hz)	= 0.078
f_9	(90 Hz)	= 0.033	f_{10}	(100 Hz)*	= 0
f_{11}	(110 Hz)	= 0.019	f_{12}	(120 Hz)	= 0.025
f_{13}	(130 Hz)	= 0.021	f_{14}	(140 Hz)	= 0.011
f_{15}	(150 Hz)*	= 0	f_{16}	(160 Hz)	= 0.008
f_{17}	(170 Hz)	= 0.011	f_{18}	(180 Hz)	= 0.01
f_{19}	(190 Hz)	= 0.006	f_{20}	(200 Hz)*	= 0

Frequencies marked * are harmonics of the 50 Hz supply.

Harmonic filters

Where harmonics present a problem on the AC system, filters can be used at the input to the converter to control their level by providing a shunt path of low impedance at the harmonic frequency. Problems in filter design include:

(a) Variation in supply (fundamental) frequency from its nominal value.
(b) Effects of ageing causing changes in filter component values and hence variation in the tuned frequency.

Fig. 7.10 Single tuned filter.

(c) Initial off-tuning as a result of manufacturing tolerances and the size of the tuning steps used.

The cost of providing filters is generally high in relation to the cost of the converter and their application tends to be confined to large converters or for the control of specific problems.

Tuned filters

The single tuned filter shown in Fig. 7.10 is a series RLC circuit which is tuned to a harmonic frequency. Its impedance is given by

$$Z_f = R + j(\omega L - 1/\omega C) \tag{7.20}$$

which reduces to R at the resonance frequency f_n. The quality factor is given by

$$Q = \omega_n L/R = 1/\omega_n CR \tag{7.21}$$

The filter passband (PB) is bounded by the frequencies f_1 and f_2 at which

$$|(\omega L - 1/\omega C)| = R \tag{7.22}$$

when

$$|Z_f| = \sqrt{2}R \tag{7.23}$$

The passband is related to the quality factor (Q) by

$$Q = \omega_n |(f_2 - f_1)| \tag{7.24}$$

where $\omega_n = 2\pi f_n$

The performance of the filter is determined largely by its quality factor and the relative frequency deviation (δ), defined by

$$\omega = \omega_n(1 + \delta) \tag{7.25}$$

where ω is the actual frequency

δ may also be expressed in terms of a change in the value of L or C as

$$\delta = \frac{\Delta f}{f} + \frac{1}{2}\left(\frac{\Delta L}{L} + \frac{\Delta C}{C}\right) \tag{7.26}$$

After manipulation of Equations 7.20, 7.21 and 7.25, the impedance of the filter can be expressed by

$$Z_f = R[1 + jQ\delta(2 + \delta)/(1 + \delta)] \tag{7.27}$$

$f_n = \omega_n/2\pi = 1/(2\pi LC)$.

Under these conditions at the upper and lower bandwidth frequencies, Z_f has a phase angle of 45°.

ω_n is the filter tuned frequency.

which, if $\delta \ll 1$, can be approximated by

$$Z_f \approx R(1 + j2\delta Q) \qquad (7.28)$$

when

$$|Z_f| \approx R(1 + 4\delta^2 Q^2)^{1/2} \qquad (7.29)$$

also

$$Y_f = 1/Z_f \qquad (7.30)$$

The harmonic voltage at the filter is then

$$V_n = I_n/(Y_n + Y_s) \qquad (7.31)$$

where Y_s is the system admittance at the point of connecting the filter

In order to estimate the maximum value of V_n likely to be encountered, the largest anticipated value of δ and the worst system admittance must be used.

The system harmonic equivalent circuit including the filter is:

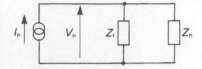

Worked example 7.5

A tuned filter with a Q of 30 is to be used to control the seventh harmonic content of a 50 Hz supply. The filter is constructed using a capacitor and inductor of tolerance $\pm 1\%$ and $\pm 2\%$ respectively. The supply frequency can vary by ± 1 Hz. What will be the ratio between the nominal impedance of the filer and the possible worst case impedance?

From Equation 7.24,

$$\delta = 0.02 + \tfrac{1}{2}(0.02 + 0.01) = 0.035$$

From Equation 7.21

$$|Z_f| = R(1 + 4 \times 0.035^2 \times 30^2)^{1/2} = 2.326R$$

Nominal impedance $= R$

Thus Ratio $= 2.326 : 1$

Automatically tuned filters

If the filters can be retuned to accommodate system changes this allows a higher Q filter to be produced, permitting the use of a lower rated capacitor. Tuning can be achieved by either switching capacitance or varying inductance. The control system operates by measuring the reactive power component in the filter at the harmonic frequency and then adjusts the tuning to reduce this component to as near zero as practical.

Damped filters

Various types of damped filter are shown in Fig. 7.11. Such filters are less sensitive to variations in frequency, component tolerances and temperature than tuned filters. In addition they provide a low impedance over a wide range of harmonics without the need for parallel branches and associated

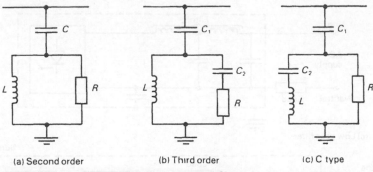

(a) Second order   (b) Third order   (c) C type

Fig. 7.11 Damped filters.

switching. Their disadvantages are that they need to be designed with higher fundamental VA ratings and the losses are generally higher than equivalent tuned filters.

Active filters

At lower power levels the use of an active filter can be considered as a means of controlling harmonic content. Essentially, an active filter consists of a secondary converter, connected in parallel with the main converter as in Fig. 7.12, whose switching is controlled to produce a near sinusoidal current in the supply.

Active filters may be either voltage sourced or current sourced and their operation is controlled by an on-board microprocessor containing the required switching algorithms.

Also referred to as a power-line conditioning circuit.

Standards

There exist a number of national and international standards, guidelines and recommendations for the control of the harmonic levels on a power system. In each case the aim is to control the system harmonic content in such a way as to provide consumers with a waveform suited to their needs and to

System standards are used to control the levels of harmonic on the transmission and distribution networks. In addition to the system standards there are several appliance and equipment standards which govern the amount of distortion that can be produced by individual items of equipment.

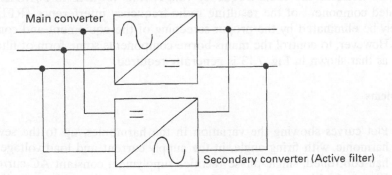

Main converter

Secondary converter (Active filter)

Fig. 7.12 Secondary converter acting as an active filter in parallel with the main converter.

197

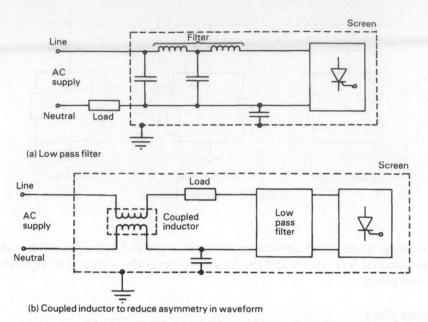

(a) Low pass filter

(b) Coupled inductor to reduce asymmetry in waveform

Fig. 7.13 Radio-frequency interference filtering.

minimize interference with system components and communications networks while allowing for the use of distortion-producing loads. The actual form of the standard is determined by the overall requirements of the power system and is therefore a function of that system. As a result the emphasis of each standard or guidelines reflects the system conditions for which it was intended, resulting in significant differences between the individual standards or guidelines. A generalized approach to harmonic control applicable over a wide range of systems is not therefore currently practical and each installation must be treated individually under the particular conditions applying.

Radio-frequency interference

The rapid transition of current that can occur when a thyristor or other power electronic device is switched gives rise to both mains-borne and radiated noise in the radio-frequency range from 150 kHz to 30 MHz. The radiated component of the resulting radio-frequency interference (RFI) can largely be eliminated by the proper screening of the equipment and conductors. However, to control the mains-borne components some form of filtering such as that shown in Fig. 7.13 is generally required.

Problems

7.1 Plot curves showing the variation in the harmonics, up to the seventh harmonic, with firing angle, in the supply current and load voltage of a half-controlled, single-phase rectifier supplying a constant AC current.

7.2 Repeat Problem 7.1 for the load voltage of a three-phase, half-controlled rectifier.

7.3 A three-phase fully-controlled converter operating at a firing angle of 30° with an overlap angle of 8° is drawing on RMS current of 16 A from a three-phase, 11 kV, 50 Hz supply busbar. Also connected to the busbar is an effective star-connected load, each phase of which consists of a series impedance at 50 Hz of $(200 + 110j)\,\Omega$ in parallel with a capacitive impedance at 50 Hz of $-840j\,\Omega$. The busbar is fed from the three-phase, 132 kV system by a 12 MVA transformer and the system fault level measured at the busbar side of the transformer is 50 MVA.

Calculate the magnitudes of the 5th, 7th, 11th and 13th harmonic currents in the 132 kV system and the corresponding harmonic voltages at the busbar.

Appendix A Three-phase systems

Phase Voltages

v, v_2 and v_3 (Fig. A.1) are measured with respect to the neutral and are the *phase* voltages:

$$v_1 = \hat{V}\sin(\omega t) \tag{A.1}$$

$$v_2 = \hat{V}\sin(\omega t - 2\pi/3) \tag{A.2}$$

$$v_3 = \hat{V}\sin(\omega t - 4\pi/3) = V\sin(\omega t + 2\pi/3) \tag{A.3}$$

Expressed as RMS phasor quantities, using v_1 as reference:

$$V_1 = V_{\text{phase}}\underline{|0°} \tag{A.4}$$

$$V_2 = V_{\text{phase}}\underline{|-120°} \tag{A.5}$$

$$V_3 = V_{\text{phase}}\underline{|-240°} = V_{\text{phase}}\underline{|120°} \tag{A.6}$$

where

$$V_{\text{phase}} = \hat{V}/\sqrt{2}$$

and is the RMS voltage.
Also

$$V_1 + V_2 + V_3 = 0 \tag{A.7}$$

Line voltages

The **line** voltages are measured between pairs of lines as shown in Fig. A.2:

$$v_{12} = v_1 - v_2 = \sqrt{3}\,\hat{V}\sin(\omega t + \pi/6) \tag{A.8}$$

$$v_{23} = v_2 - v_3 = \sqrt{3}\,\hat{V}\sin(\omega t - \pi/2) \tag{A.9}$$

$$v_{31} = v_3 - v_1 = \sqrt{3}\,\hat{V}\sin(\omega t - 7\pi/6) = 3\sqrt{}\sin(\omega t + 5\pi/6) \tag{A.10}$$

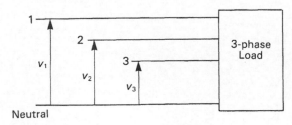

Fig. A.1. Phase voltages

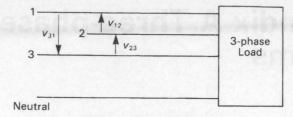

Fig. A.2. Line voltages

Expressed as **phasor** quantities, using v_1 as reference:

$$V_{12} = V_1 - V_2 = \sqrt{3}V_{\text{phase}}\underline{|30°} = V_{\text{line}}\underline{|30°} \qquad (A.11)$$

$$V_{23} = V_2 - V_3 = \sqrt{3}V_{\text{phase}}\underline{|-90°} = V_{\text{line}}\underline{|-90°} \qquad (A.12)$$

$$V_{31} = V_3 - V_1 = \sqrt{3}V_{\text{phase}}\underline{|-210°} = \sqrt{3}V_{\text{phase}}\underline{|150°} = V_{\text{line}}\underline{|150°} \qquad (A.13)$$

Also

$$V_{12} + V_{23} + V_{31} = 0 \qquad (A.14)$$

and

$$|V_{\text{line}}| = \sqrt{3}|V_{\text{phase}}| \qquad (A.15)$$

Appendix B Fourier analysis

The Fourier series of a periodic function $x(t)$ of period T has the form

$$x(t) = a_0 + \sum_{n=1}^{\infty} \{a_n \cos(2\pi nt/T) + b_n \sin(2\pi nt/T)\} \tag{B.1}$$

where a_0 is the mean value of the function $x(t)$, while a_n and b_n, the series coefficients are the rectangular components of the nth harmonic.

The magnitude of the nth harmonic is

$$A_n = (a_n^2 + b_n^2)^{1/2} \tag{B.2}$$

at a phase angle of

$$\phi_n = \tan^{-1}(b_n/a_n) \tag{B.3}$$

a_0, a_n and b_n are calculated from

$$a_0 = \frac{1}{T} \int_{-T/2}^{T/2} x(t)\, dt = \frac{1}{2\pi} \int_{-\pi}^{\pi} x(\omega t)\, d(\omega t) \tag{B.4}$$

$$a_n = \frac{2}{T} \int_{-T/2}^{T/2} x(t) \cos(2\pi nt/T)\, dt = \frac{1}{\pi} \int_{-\pi}^{\pi} x(\omega t) \cos(n\omega t)\, d(\omega t) \tag{B.5}$$

$$b_n = \frac{2}{T} \int_{-T/2}^{T/2} x(t) \sin(2\pi nt/T)\, dt = \frac{1}{\pi} \int_{-\pi}^{\pi} x(\omega t) \sin(n\omega t)\, d(\omega t) \tag{B.6}$$

The above may be simplified by the use of waveform symmetry.

Odd symmetry

The waveform has odd symmetry when

$$x(t) = -x(-t) \tag{B.7}$$

in which case the a_n coefficients have a value of zero for all n and the Fourier series will contain only sine terms.

Even symmetry

The waveform has even symmetry when

$$x(t) = x(-t) \tag{B.8}$$

in which case the b_n coefficients have a value of zero for all n and the Fourier series will contain only cosine terms.

Half-wave symmetry

The function has half-wave symmetry when

$$x(t) = -x(t + t/2) \tag{B.9}$$

in which case the Fourier series will contain odd order harmonics only.

Reference

Arrillaga, J., Bradley, D.A. and Bodger, P.S. (1985). *Power System Harmonics*. John Wiley, U.K.

Answers to problems

1.1 (a) 44.41 W; 36.78 A
    (b) 9.45 W; 12.7 A
    (c) 46.82 W; 35.96 A

1.2 Turn-on t (μs)

5	10	15	20	25	30	35	40

Power (W)

840	1800	1920	1800	1440	840	0

Turn-off t (μs)

5	10	15	20	25	30	35	40	45	50	55

Power (W)

586.7	1066.7	1440	1706	1866.7	1920	1866.7	1706	1440	1066.7	586.7

1.3 $1.94°C\,W^{-1}$; 104°C

1.4 31°C

1.5 (a) $I_1 = 363\,A$; $I_2 = 37\,A$
    (b) $I_1 = 558\,A$; $I_2 = 242\,A$
    (c) $I_1 = 752.9\,A$; $I_2 = 447.1\,A$
    (d) $I_1 = 947.9\,A$; $I_2 = 652.1\,A$
    (e) $I_1 = 1142.9\,A$; $I_2 = 857.1\,A$
    $R = 0.445\,m\Omega$;
    400 A load; $I_1 = 256.8\,A$; $I_2 = 143.2\,A$
    2000 A load; $I_1 = 1049.8\,A$; $I_2 = 950.2\,A$

2.1 For a resistive load current is in phase with voltage and the conduction angle is
    $(180 - \alpha)°$
    $\alpha = 30°$; $I_{mean} = 6.16\,A$; $I_{max} = 20.74\,A$; $I_{RMS} = 10.22\,A$
    $\alpha = 60°$; $I_{mean} = 4.95\,A$; $I_{max} = 20.74\,A$; $I_{RMS} = 10.07\,A$
    Firing angle $= 20° 39'$

2.2 $\alpha = 0°$; Power $= 11.2\,kW$
    $\alpha = 75°$; Power $= 2.9\,kW$

2.3 The waveform will be as Fig. 2.3(b)

2.4 Turn-on, overlap $= 0.9°$
    Turn-off, overlap $= 9.5°$

2.5 Primary rating $= 23.8\,kW$; Secondary rating $= 33.7\,kW$

2.6 $\alpha = 63° 10'$; $I_{RMS} = 34.64\,A$; Power $= 24\,W$
    $\alpha = 58° 23'$

2.7 Thyristor rating, Parallel $= 144.33\,A$; Series $= 288.67\,A$

2.8 Voltage $= 506.6\,V$; $\gamma = 8° 17'$; $\delta = 21° 43'$

2.9 $I = 67.53\,A$

3.1 (a) 189.3 W; (b) 134.4 W

3.2 (a) $V = 55\,V$; $I = 11\,A$
    (b) $V = 91.67\,V$; $I = 18.33\,A$
    (c) $V = 27.5\,V$; $I = 5.5\,A$

3.3 Mark–space ratio = 60%; $I_{\text{mean}} = 9\,\text{A}$; $I_1 - I_2 = 0.3\,\text{A}$

3.4 (a) 5829 W (b) 3352 W

3.5 $L = 12.5\,\text{H}$, $C = 1.8\,\mu\text{F}$

3.6 $V = 988\,\text{V}$ (line)

Peak current = 101.8 A

Thyristor RMS current = 58.8 A

Load power = 29.4 kW; Input power factor = 0.41

3.7 (a) $I_{\text{RMS}} = 16.33\,\text{A}$; Power = 14.4 kW; Rating = 11.5 A

(a) $I_{\text{RMS}} = 18.85\,\text{A}$; Power = 19.2 kW; Rating = 13.33 A

4.1 237 rev min^{-1}

4.2 14.86 N m

4.3 13.08 N m

4.4 See notes on pages 126 and 127

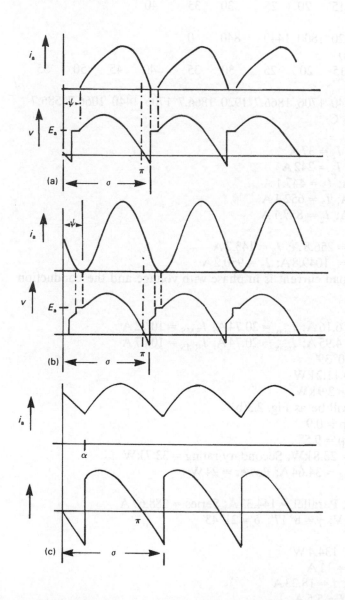

4.5 (a) $1267\,\mathrm{rev\,min^{-1}}$; (b) $1599\,\mathrm{rev\,min^{-1}}$; (c) $107\,\mathrm{N\,m}$; (d) $56.7\,\mathrm{N\,m}$

4.6 Copper $1000 = 266.8\,\mathrm{W}$
$P_{ag} = 5717\,\mathrm{W}$

4.7 $600\,\mathrm{rev\,min^{-1}} \equiv 20\,\mathrm{Hz}$ and $166\,\mathrm{V}$
$900\,\mathrm{rev\,min^{-1}} \equiv 30\,\mathrm{Hz}$ and $249\,\mathrm{V}$
$1200\,\mathrm{rev\,min^{-1}} \equiv 40\,\mathrm{Hz}$ and $332\,\mathrm{V}$

4.8 $1296\,\mathrm{rev\,min^{-1}}$ and $533.4\,\mathrm{N\,m}$

5.1 Peak transistor current $= 311\,\mathrm{mA}$, $L = 392.7\,\mu\mathrm{H}$
$C_o = 5.7\,\mu\mathrm{F}$
C_o needs to be increased to allow for capacitor equivalent resistance

5.2 Peak transistor current $= 341\,\mathrm{mA}$, $L = 339.8\,\mu\mathrm{H}$
$C_o = 5.89\,\mu\mathrm{F}$
C_o needs to be increased as before

6.1 $I_{RMS} = 1804\,\mathrm{A}$; Peak reverse voltage $= 167.6\,\mathrm{kV}$

6.2 Rating (RMS) $= 33.39\,\mathrm{A}$
@ $16\,\mathrm{kW}$, $\alpha = 74.49°$
@ $8\,\mathrm{kW}$, $\alpha = 105.29°$
Thyristor rating (RMS) $= 23.61\,\mathrm{A}$

6.3 (a) $3441\,\mathrm{W}$; (b) $1370\,\mathrm{W}$; (c) $177.2\,\mathrm{W}$

6.4 $V_L = V_{max} \left(\dfrac{1}{\pi} \left[\dfrac{\pi}{2} - \dfrac{5\alpha}{18} + \dfrac{10}{18} \sin 2\alpha \right] \right)^{1/2}$

7.1 Voltage
$a_n = 0$ for n odd
$= \dfrac{V_m}{\pi} \left[\dfrac{2}{1-n^2} + \dfrac{\cos[(1+n)\alpha]}{1+n} + \dfrac{\cos[(1-n)\alpha]}{1-n} \right] n$ even

$b_n = 0$ for n odd
$= \dfrac{V_m}{\pi} \left[\dfrac{\sin[(1+n)\alpha]}{1+n} - \dfrac{\sin[(1-n)\alpha]}{1-n} \right] n$ even

Current
$a_n = 0$ for n even
$= -\dfrac{2}{n\pi} \sin(n\alpha)$ for n odd

$b_n = 0$ for n even
$= \dfrac{2}{n\pi}(1 + \cos(n\alpha))$ for n odd

7.2 $a_n = \dfrac{3}{2\pi} \left[\dfrac{\sin[(3n+1)\alpha]}{3n+1} + \dfrac{\sin[(1-3n)\alpha]}{1-3n} - \dfrac{\sqrt{3}\cos(n\pi)}{1-9n^2} \right]$

$b_n = \dfrac{3}{2\pi} \left[\dfrac{\cos[(1+3n)\alpha]}{1+3n} + \dfrac{\cos[(1-3n)\alpha]}{1-3n} - \dfrac{\cos(n\pi)}{1-9n^2} \right]$

7.3 $I_5 = 0.277\,\mathrm{A}$
$I_7 = 0.22\,\mathrm{A}$
$I_{11} = 0.238\,\mathrm{A}$
$I_{13} = 0.459\,\mathrm{A}$

Index

213